The Constructive Systems Engineering Cost Model (COSYSMO)

Ricardo Valerdi

The Constructive Systems Engineering Cost Model (COSYSMO)

Quantifying the Costs of Systems Engineering
Effort in Complex Systems

VDM Verlag Dr. Müller

Imprint

Bibliographic information by the German National Library: The German National Library lists this publication at the German National Bibliography; detailed bibliographic information is available on the Internet at http://dnb.d-nb.de.

Cover image: www.purestockx.com

Publisher:
VDM Verlag Dr. Müller Aktiengesellschaft & Co. KG , Dudweiler Landstr. 125 a, 66123 Saarbrücken, Germany,
Phone +49 681 9100-698, Fax +49 681 9100-988,
Email: info@vdm-verlag.de

Zugl.: Los Angeles, University of Southern California, Diss., 2005

Produced in USA and UK by:
Lightning Source Inc., La Vergne, Tennessee, USA
Lightning Source UK Ltd., Milton Keynes, UK
BookSurge LLC, 5341 Dorchester Road, Suite 16, North Charleston, SC 29418, USA

ISBN: 978-3-639-03478-3

DEDICATION

This book is dedicated to my mother and father,
Lucía and Jorge.

ACKNOWLEDGEMENTS

If I have been able to see further than others, it is because I have stood on the shoulders of giants. -- Sir Isaac Newton

No intellectual achievement occurs in a vacuum. All new creativity builds on the efforts that have gone before. Like Newton, I have been able to stand on the shoulders of extremely talented people. I am forever in debt to these giants which have contributed intellectual ingredients to this work. First, my family for providing a strong foundation. Second, my academic advisors and colleagues for exposing me to the world of engineering. And third, the organizations that supported this research through funding, expertise, and data.

My mother, Lucia, for teaching me values that have helped me become a member of society and my father for teaching me how to use those values to make a contribution. To my fiancée Briana for her unconditional support and unending patience.

The ideas presented here exist as a result of the trailblazing vision and persistence of my advisor, Dr. Barry W. Boehm. Unconditional intellectual support was provided by Dr. Stan Settles, Dr. George Friedman, Dr. Ann Majchrzak, Dr. Elliot Axelband, Dr. Bert Steece and Don Reifer.

The realization of this model exists because of the tremendous support of the Center for Systems & Software Engineering corporate affiliates. Specifically, Gary Thomas from Raytheon whose development of *myCOSYSMO* served as a catalyst for the acceptance of the model among practitioner circles. Special thanks to Merrill Palmer, John Gaffney, and Dan Ligett for thoroughly reviewing this manuscript. Others providing intellectual support are listed in Appendix C.

I am grateful for the support of Marilee Wheaton and Pat Maloney from The Aerospace Corporation. Additional support was provided by the Air Force Space and Missile Systems Center, Office of the Chief Engineer. This research has also received visibility and endorsement from: the International Council on Systems Engineering (Corporate Advisory Board, Measurement Working Group, and Systems Engineering Center of Excellence); Southern California Chapter of the International Society of Parametric Analysts; Practical Software & Systems Measurement; and the Space Systems Cost Analysis Group.

TABLE OF CONTENTS

LIST OF TABLES

3

4

5

LIST OF FIGURES

ABREVIATIONS

ANSI	American National Standards Institute
C4ISR	Command Control Communications Computer Intelligence Surveillance Reconnaissance
CER	Cost Estimation Relationship
CM	Configuration Management
CMM	Capability Maturity Model
CMMI	Capability Maturity Model Integration
COCOMO II	Constructive Cost Model version II
COCOTS	Constructive Commercial-off-the-shelf Model
COPROMO	Constructive Productivity Model
COPSEMO	Constructive Phased Schedule Estimation Model
COQUALMO	Constructive Quality Model
CORADMO	Constructive Rapid Application Development Model
COSOSIMO	Constructive System-of-systems Cost Model
COSYSMO	Constructive Systems Engineering Cost Model
CSSE	Center for Systems & Software Engineering
CSER	Conference on Systems Engineering Research
DCAA	Defense Contract Audit Agency
DF	Degrees of Freedom
DoD	Department of Defense
EIA	Electronic Industries Alliance
EM	Effort Multiplier
EMR	Effort Multiplier Ratio
GAO	Government Accountability Office
GUTSE	Grand Unified Theory of Systems Engineering
IEC	International Electrotechnical Commission
IEEE	Institute of Electrical and Electronics Engineers
IKIWISI	I'll Know It When I See It
INCOSE	International Council on Systems Engineering
IP	Information Processing
ISO	International Organization for Standardization
KPA	Key Process Area
KPP	Key Performance Parameter
KSLOC	Thousands of Software Lines of Code
MBASE	Model Based System Architecting and Software Engineering
MIL-STD	Military Standard
MITRE	MIT Research Corporation
MMRE	Mean Magnitude of Relative Error
MSE	Mean Square Error
OLS	Ordinary Least Squares
OTS	Off The Shelf
PM	Person Month
PRED	Prediction level
PRICE	Parametric Review of Information for Costing and Evaluation
RSERFT	Raytheon Systems Engineering Resource Forecasting Tool

RSS	Residual Sum of Squares
RUP	Rational Unified Process
SE	Systems Engineering
SEER	System Evaluation and Estimation of Resources
SEMP	Systems Engineering Management Plan
SMC	Space and Missile Systems Center
SoS	System-of-systems
SSCM	Small Satellite Cost Model
SW	Software
TPM	Technical Performance Measure
TRL	Technology Readiness Level
USCM	Unmanned Satellite Cost Model
WBS	Work Breakdown Structure

1. Introduction

1.1. Motivation for a Systems Engineering Cost Model

It is clear that we have been living in the Systems Age for some time as evidenced by the role of technologically enabled systems in our every day lives. Most of our every day functions are dependent on, or enabled by, large scale man made systems that provide useful technological capabilities. The advent of these systems has created the need for systems thinking and ultimately systems engineering.

The function of systems engineering – coupled with the other traditional disciplines such as electrical engineering, mechanical engineering, or civil engineering – enables the creation and implementation of systems of unprecedented size and complexity. However, these disciplines differ in the way they create value. Traditional engineering disciplines are value-neutral; the laws of physics control the outcome of electronics, mechanics, and structures. Tangible products serve as evidence of the contribution that is easily quantifiable. Systems engineering has a different paradigm in that its intellectual output is often intangible and more difficult to quantify. Common work artifacts such as requirements, architecting, design, verification, and validation are not readily noticed. For this reason, systems engineering is better suited for value-based approach artifacts where value considerations are integrated with systems engineering principles and practices. The link between systems engineering artifacts to cost and schedule is recognized but currently not well understood. This leads to the principal research question addressed in this book:

How much systems engineering effort, in terms of person months, should be allocated for the successful conceptualization, development, and testing of large-scale systems?

The model presented in this book, COSYSMO, helps address this issue using a value-based approach.

1.1.1. Fundamentals of Systems Engineering

Systems engineering is concerned with creating and executing an interdisciplinary process to ensure that the customer and stakeholder needs are satisfied in a high quality, trustworthy, cost efficient and schedule compliant manner throughout a system's entire life cycle. Part of the complexity in understanding the cost involved with systems engineering is due to the diversity of definitions used by different systems engineers and the unique ways in which systems engineering is used in practice. The premier systems engineering society,

9

INCOSE, has long debated the definition of systems engineering and only recently converged on the following:

> *Systems Engineering is an interdisciplinary approach and means to enable the realization of successful systems. It focuses on defining customer needs and required functionality early in the development cycle, documenting requirements, then proceeding with design synthesis and system validation while considering the complete problem.*

Experts have provided their own definitions of systems engineering as shown in Table 1.

Table 1 Collection of Definitions of Systems Engineering

Source	Definition
Simon Ramo (Jackson 2002)	*A branch of engineering that concentrates on the design and application of the whole as distinct from the parts...looking at a problem in its entirety, taking into account all the facets and all the variables and relating the social to the technical aspects.*
George Friedman (Jackson 2002)	*That engineering discipline governing the design, evaluation, and management of a complex set of interacting components to achieve a given purpose [function].*
Andrew Sage (Sage 1992)	*Systems engineering involves the application of a general set of guidelines and methods useful for assisting clients in the resolution of issues and problems [system definition] which are often large scale and scope. Three fundamental steps may be distinguished: (a) problem or issue formulation [requirements], (b) problem or issue analysis [synthesis] and (c) interpretation of analysis results [verification].*
Ben Blanchard and Wolter Fabrycky, (Blanchard and Fabrycky 1998)	*The application of efforts necessary to (1) transform an operational need into a description of system performance [requirements] (2) integrate technical parameters and assure compatibility of all physical, functional and program interfaces in a manner that optimizes [or balances] the total system definition and design [synthesis] and (3) integrate performance, producibility, reliability, maintainability, manability [human operability], supportability and other specialties into the total engineering effort.*

Each of these definitions are appropriate for different situations. Each of them contains a different perspective that is representative of the application of the principles of systems engineering. These definitions also highlight the broad applicability of systems engineering across domains. Defining systems engineering is the first step in understanding it. Managing it, however, requires a deeper understanding of the cost and tradeoffs associated with it.

10

A constituency of practitioners familiar with the benefits provided by the Constructive Cost Model (COCOMO) in the realm of software engineering proposed the development of a similar model to focus on systems engineering (Boehm, Egyed et al. 1998). No formal approach to estimating systems engineering existed at the time, partially because of the immaturity of systems engineering as a formal discipline and the lack of mature metrics. The beginnings of systems engineering can be traced back to the Bell Telephone Laboratories in the 1940s (Auyang 2004). However, it was not until almost thirty years later that the first U.S. military standard was published (MIL-STD-499A 1969). The first professional systems engineering society, INCOSE, was not organized until the early 1990s and the first commercial U.S. systems engineering standards, ANSI/EIA 632 and IEEE 1220, followed shortly thereafter. Even with the different approaches of defining systems engineering, the capability to estimate it is desperately needed by organizations. Several heuristics are available but they do not provide the necessary level of detail that is required to understand the most influential factors and their sensitivity to cost.

Fueled by industry support and the US Air Force's systems engineering revitalization initiative (Humel 2003), interest in COSYSMO has grown. Defense contractors as well as the federal government are in need of a model that will help them better control and prevent future shortfalls in the $18 billion federal space acquisition process (GAO 2003). COSYSMO is also positioned to make immediate impact on the way organizations – and other engineering disciplines – view systems engineering.

Based on the previous support for COCOMO II, COSYSMO is positioned to leverage off the existing body of knowledge developed by the software community. The synergy between software engineering and systems engineering is intuitive because of the strong linkages in their products and processes. Researchers identified strong relationships between the two disciplines (Boehm, 1994), opportunities for harmonization (Faisandier & Lake, 2004), and lessons learned (Honour, 2004). There have also been strong movements towards convergence between software and systems as reflected in two influential standards: ISO 15504 *Information technology - Process assessment* and the CMMI[1]. Organizations are going as far as changing their names to reflect their commitment and interest in this convergence. Some examples include the Software Productivity Consortium becoming the Systems & Software Consortium and the Software Technology Conference becoming the

---

[1] Capability Maturity Model Integration

Software & Systems Technology Conference. Despite the strong coupling between software and systems they remain very different activities in terms of maturity, intellectual advancement, and influences regarding cost.

1.1.2. Comparison Between COCOMO II and COSYSMO

On the surface, COCOMO II and COSYSMO appear to be similar. However, there are fundamental differences between them that should be highlighted. These are obvious when the main assumptions of the model are considered:

- Sizing. COCOMO II uses software size metrics while COSYSMO uses metrics at a level of the system that incorporates both hardware and software.

- Life cycle. COCOMO II, based on a software tradition, focuses exclusively on software development life cycle phases defined by MBASE[2] (Boehm and Port 1999) while COSYSMO follows the system life cycle provided by ISO/IEC 15288.

- Cost Drivers. Each model includes drivers that model different phenomena. The overlap between the two models is minimal since very few of the COCOMO II parameters are applicable to systems engineering. One appreciable overlap is the software-related systems engineering effort estimated by both models. This overlap is covered in section 4.2

A more fundamental difference between the two models is that COCOMO II benefits from existing software engineering metrics. COSYSMO does not benefit from such a deep body of knowledge. As the first model to focus on issues outside of the software domain, it faces numerous challenges.

---

[2] Model Based System Architecting and Software Engineering

Table 2 Differences between COCOMO II and COSYSMO

	COCOMO II	COSYSMO
Estimates	Software development	Systems engineering
Estimates size via	Thousands of Software Lines of Code (KSLOC), Function Points, or Application Points	Requirements, Interfaces, Algorithms, and Operational Scenarios
Life cycle phases	MBASE/RUP Phases: (1) Inception, (2) elaboration, (3) construction, and (4) transition	ISO/IEC 15288 Phases: (1) Conceptualize, (2) Develop, (3) Operation, Test, and Evaluation, (4) Transition to Operation, (5) Operate Maintain or Enhance, and (6) Replace or dismantle.
Form of the model	1 size factor, 5 scale factors, and 18 effort multipliers	4 size factors, 1 scale factor, 14 effort multiplier
Represents diseconomy of scale through	Five scale factors	One exponential system factor

COCOMO II was a natural starting point which provided a useful and mature framework. The scope of this book is to identify the relevant parameters in systems engineering while building from the lessons learned in software cost estimation. As much synergy as exists, software engineering and systems engineering must be treated as independent activities. This involves measuring them independently and identifying metrics that best capture the size and cost factors for each.

1.1.3. COSYSMO Objectives

COSYSMO is a model that can help people reason about the cost implications of systems engineering. User objectives include the ability to make the following:

- Investment decisions. A return-on-investment analysis involving a systems engineering effort needs an estimate of the systems engineering cost or a life cycle effort expenditure profile.

- Budget planning. Managers need tools to help them allocate project resources.

- Tradeoffs. Decisions often need to be made between cost, schedule, and performance.

- Risk management. Unavoidable uncertainties exist for many of the factors that influence systems engineering.

13

- Strategy planning. Setting mixed investment strategies to improve an organization's systems engineering capability via reuse, tools, process maturity, or other initiatives.

- Process improvement measurement. Investment in training and initiatives often need to be measured. Quantitative management of these programs can help monitor progress.

To enable these user objectives the model has been developed to provide certain features to allow for decision support capabilities. Among these is to provide a model that is:

- Accurate. Where estimates are close to the actual costs expended on the project. See section 5.2.1.

- Tailorable. To enable ways for individual organizations to adjust the model so that it reflects their business practices. See section 5.2.4.

- Simple. Understandable counting rules for the drivers and rating scales. See section 3.2.

- Well-defined. Scope of included and excluded activities is clear. See sections 3.2 and 3.3.

- Constructive. To a point that users can tell why the model gives the result it does and helps them understand the systems engineering job to be done.

- Parsimonious. To avoid use of highly redundant factors or factors which make no appreciable contribution to the results. See section 5.2.2.

- Pragmatic. Where inputs to the model correspond to the information available early on in the project life cycle.

This research puts these objectives into context with the exploration of what systems engineering means in practice. Industry standards are representative of collective experiences that help shape the field as well as the scope of COSYSMO.

1.2. Systems Engineering and Industry Standards

The synergy between software engineering and systems engineering is evident by the integration of the methods and processes developed by one discipline into the culture of the other. Researchers from software engineering (Boehm 1994) and systems engineering (Rechtin 1998) have extensively promoted the integration of both disciplines but have faced roadblocks that result from the fundamental difference between the two disciplines (Pandikow and Törne 2001).

14

The development of systems engineering standards has helped the crystallization of the discipline as well as the development of COSYSMO. Table 3 includes a list of the standards most influential to this effort.

Table 3 Notable Systems Engineering Standards

Standard (year)	Title
MIL-STD-499A (1969)	Engineering Management
MIL-STD-490-A (1985)	Specification Practices
ANSI/EIA-632 (1999)	Processes for Engineering a System
CMMI (2002)	Capability Maturity Model Integration
ANSI/EIA-731.1 (2002)	Systems Engineering Capability Model
ISO/IEC 15288 (2002)	Systems Engineering – System Life Cycle Processes

The first U.S. military standard focused on systems engineering provided the first definition of the scope of engineering management (MIL-STD-499A 1969). It was followed by another standard that provided guidance on the process of writing system specifications for military systems (MIL-STD-490A 1985). These standards were influential in defining the scope of systems engineering in their time. Years later the standard ANSI/EIA 632 *Processes for Engineering a System* (ANSI/EIA 1999) provided a typical systems engineering WBS[3]. This list of activities was selected as the baseline for defining systems engineering in COSYSMO. The standard contains five fundamental processes and 13 high level process categories that are representative of systems engineering organizations. The process categories are further divided into 33 activities shown in Appendix A. These activities help answer the *what* of systems engineering and helped characterize the first significant deviation from the software domain covered by COCOMO II. The five fundamental processes are (1) Acquisition and Supply, (2) Technical Management, (3) System Design, (4) Product Realization, and (5) Technical Evaluation. These processes are the basis of the systems engineering effort profile developed for COSYSMO. The effort profile is provided in Appendix B.

This standard provides a generic industry list which may not be applicable to every situation. Other types of systems engineering WBS lists exist such as the one developed by Raytheon Space & Airborne Systems (Ernstoff and Vincenzini 1999). Lists such as this one provide, in much finer detail, the common activities that are likely to be performed by systems engineers in those organizations, but are generally not applicable outside of the companies or application domains in which they are created.

---

[3] Work Breakdown Structure

Under the integrated software engineering and systems engineering paradigm, or Capability Maturity Model Integration® (CMMI 2002), software and systems are intertwined. A project's requirements, architecture, and process are collaboratively developed by integrated teams based on shared vision and negotiated stakeholder concurrence. A close examination of CMMI process areas – particularly the staged representation – strongly suggests the need for the systems engineering function to estimate systems engineering effort and cost as early as CMMI Maturity Level 2. Estimates can be based upon a consistently provided organizational approach from past project performance measures related to size, effort and complexity. While it might be possible to achieve high CMMI levels without a parametric model, an organization should consider the effectiveness and cost of achieving them using other methods that may not provide the same level of stakeholder confidence and predictability. The more mature an organization, the more benefits in productivity they experience (ANSI/EIA 2002).

After defining the possible systems engineering activities used in COSYSMO, a definition of the system life cycle phases is needed to help define the model boundaries. Because the focus of COSYSMO is systems engineering, it employs some of the life cycle phases from ISO/IEC 15288 *Systems Engineering – System Life Cycle Processes* (ISO/IEC 2002). These phases were slightly modified to reflect the influence of the aforementioned model, ANSI/EIA 632, and are shown in Figure 1.

Figure 1 COSYSMO System Life Cycle Phases

Life cycle models vary according to the nature, purpose, use and prevailing circumstances of the system. Despite an infinite variety in system life cycle models, there is an essential set of characteristic life cycle phases that exists for use in the systems engineering domain. For example, the *Conceptualize* stage focuses on identifying stakeholder needs, exploring different solution concepts, and proposing candidate solutions. The *Development* stage involves refining the system requirements, creating a solution description, and building a system. The *Operational Test & Evaluation* stage involves

16

verifying/validating the system and performing the appropriate inspections before it is delivered to the user. The *Transition to Operation* stage involves the transition to utilization of the system to satisfy the users' needs. These four life cycle phases are within the scope of COSYSMO. The final two were included in the data collection effort but did not yield enough data to perform a calibration. These phases are: *Operate, Maintain, or Enhance* which involves the actual operation and maintenance of the system required to sustain system capability, and *Replace or Dismantle* which involves the retirement, storage, or disposal of the system.

Each stage has a distinct purpose and contribution to the whole life cycle and represents the major life cycle periods associated with a system. The stages also describe the major progress and achievement milestones of the system through its life cycle. These life cycle stages help answer the *when* of systems engineering and COSYSMO. Understanding when systems engineering is performed relative to the system life cycle helps define anchor points for the model.

System-of-Interest. The ISO/IEC 15288 standard also provides a structure that helps define the system hierarchy. Systems can be characterized by their architectural structure or levels of responsibility. Each project has the responsibility for using levels of system composition beneath it and creating an aggregate system that meets the customer's requirements. Each particular subproject views its system as a system-of-interest within the grand scheme. The subproject's only task may be to deliver their system-of-interest to a higher level in the hierarchy. The top level of the hierarchy is then responsible for integrating the subcomponents that are delivered and providing a functional system. Essential services or functionalities are required from the systems that make up the system hierarchy. These systems, called enabling systems, can be made by the organization itself or purchased from other organizations.

The system-of-interest framework helps answer the *where* of systems engineering for use in COSYSMO. In the case where systems engineering takes place at different levels of the hierarchy, organizations should focus on the portion of the system which they are responsible for testing. Identifying system test responsibility helps crystallize the scope of the systems engineering estimate at a specific level of the system hierarchy.

The diversity of systems engineering standards can be quite complex (Sheard 1997), therefore only the applicable standards have been mentioned here. With the need and general context for the model defined, the central proposition and hypotheses for this research are proposed.

1.3. Proposition and Hypotheses

Clear definitions of the *what, when, and where* of systems engineering sets the stage for the statement of purpose for COSYSMO. The central proposition at the core of this research is:

> *There exists a subset of systems engineering projects for which it is possible to create a parametric model that will estimate systems engineering effort*
>
> *(a) for specific life cycle phases*
>
> *(b) at a certain level of system decomposition*
>
> *(c) with the same statistical criteria as the COCOMO suite of models at a comparable stage of maturity in time and effort*

This statement provides the underlying goal of the model by clarifying its solution space. The selection of the *subset of systems engineering projects* attempts to provide a homogenous group of projects from which the model can be based. For the COSYSMO data set, useful discriminators included: systems engineering productivity, systems engineering domain, and organization providing the data. The term *parametric* implies that a given equation represents a mean function that is characteristic of Cost Estimating Relationships in systems engineering. Specific *life cycle phases* are selected based on the data provided by industry participants. Counting rules are provided for a *level of system decomposition* to ensure uniform counting rules across organizations that use the model. Similar *statistical criteria* are used to evaluate COSYSMO for comparison with other cost estimation models.

The central proposition was validated through the use of the scientific method (Isaac and Michael 1997) and analysis of data (Cook and Weisberg 1999) with the aim of developing a meaningful solution. In terms of scientific inquiry, the model was validated through the following hypotheses:

> *H#1: A combination of the four elements of functional size in COSYSMO contributes significantly to the accurate estimation of systems engineering effort.*

The criteria used was a significance level less than or equal to 0.10 which translates to a 90% confidence level that these elements are significant.

> *H#2: An ensemble of COSYSMO effort multipliers contribute significantly to the accurate estimation of systems engineering.*

18

The same significance level of 0.10 was used to test this hypothesis.

H#3: The value of the COSYSMO exponent, E, which can represent economies/diseconomies of scale is greater than 1.0.

To test this hypothesis, different values for E were calculated and their effects were tested on model accuracy.

H#4: There exists a subset of systems engineering projects for which it is possible to create a parametric model that will estimate systems engineering effort at a PRED(30) accuracy of 50%.

Various approaches were used to fine tune the model and bring to a point where it was possible to test this hypothesis.

Each hypothesis is designed to test key assumptions of the model. These assumptions, as well as the structure of the model, are discussed in more detail in the next section. In addition to the four quantitative hypotheses, a qualitative hypothesis was developed to test the impact of the model on organizations.

H#5: COSYSMO makes organizations think differently about Systems Engineering cost.

The hypothesis was validated through interviews with engineers from the participating companies that provided historical data and expert opinion in the Delphi survey.

2. Background and Related Work

2.1. State of the Practice

The origins of parametric cost estimating date back to World War II (NASA 2002). The war caused a demand for military aircraft in numbers and models that far exceeded anything the aircraft industry had manufactured before. While there had been some rudimentary work to develop parametric techniques for predicting cost, there was no widespread use of any cost estimating technique beyond a bottoms-up buildup of labor-hours and materials. A type of statistical estimating was suggested in 1936 by T. P. Wright in the Journal of Aeronautical Science. Wright provided equations which could be used to predict the cost of airplanes over long production runs, a theory which came to be called the learning curve. By the time the demand for airplanes had exploded in the early years of World War II, industrial engineers were using Wright's learning curve to predict the unit cost of airplanes. Today, parametric cost models are used for estimating software development (Boehm, Abts et al. 2000), unmanned satellites (USCM 2002), and hardware development (PRICE-H 2002).

A parametric cost model is defined as: a group of cost estimating relationships used together to estimate entire cost proposals or significant portions thereof. These models are often computerized and may include many interrelated Cost Estimation Relationships (CERs), both cost-to-cost and cost-to-non-cost. The use of parametric models in engineering management serves as valuable tools for engineers and project managers to estimate engineering effort. Developing these estimates requires a strong understanding of the factors that affect, in this case, systems engineering effort.

An important part of developing a model such as COSYSMO is recognizing previous work in related areas. This process often provides a stronger case for the existence of the model and ensures that its capabilities and limitations are clearly defined. This section provides an overview of an analysis done on eight existing cost models - three of which focus on software and five on hardware (Valerdi, Ernstoff et al. 2003). These models include SE components and each employs its own unique approaches to sizing systems. An overview of the genesis and assumptions of each model sheds light on their individual applicability. While it has been shown that the appropriate level of SE effort leads to better control of project costs (Honour 2002), identifying the necessary level of SE effort is not yet a mature process. Some projects use the traditional 15% of the prime mission product or prime mission equipment to estimate systems engineering, while other projects tend to use informal

20

rules of thumb. These simplified and inaccurate methods can lead to excessively high bids by allocating too many hours on SE or, even worse, may underestimate the amount of SE needed.

One significant finding during the review was that SE costs were extremely sensitive to the sizing rules that formed the basis of these models. These rules help estimators determine the functional size of systems and, by association, the size of the job. Similar comparative analysis of cost models has been completed (Kemerer 1987), which focused exclusively on models for software development. Going one step further, both software and hardware cost models are considered since they are both tightly coupled with SE.

Cost models have been an essential part of DoD acquisition since the 1970s. Hardware models were the first to be developed and were followed by software models in the 1980s (Ferens 1999). The corresponding owner/developer and domain of applicability for the models of interest is provided in Table 4.

Table 4 Cost Models With Systems Engineering Components

Model Name	Owner/Developer	Domain
COCOMO II	USC	Software
PRICE-H	PRICE Systems, LLC	Hardware
PRICE-S	PRICE Systems, LLC	Software
Raytheon SE Resource Forecasting Tool	Raytheon	Hardware
SEER-H	Galorath, Inc.	Hardware
SEER-SEM	Galorath, Inc.	Software
SSCM	The Aerospace Corporation	Hardware
USCM8	Los Angeles Air Force Base	Hardware

The eight aforementioned models were compared in five key areas relevant to systems engineering:

1. Model inputs for software or hardware size
2. Definition of systems engineering
3. Model inputs for systems engineering
4. Life Cycle stages used in the model
5. Domain of applicability

These areas provided valuable information on the applicability of each model to systems engineering sizing. The increasing frequency and number of programs that have run significantly over-budget and behind schedule (GAO-03-1073 2003) because SE problems were not adequately understood should, by itself, be reason enough for the acquisition community to press for improvement in forecasting SE resource needs. However, even if the history of SE problems is ignored, the future paints an even more demanding picture. The undeniable trend is toward increasingly complex systems dependent on the coordination of

21

interdisciplinary developments where effective system engineering is no longer just another technology, but the key to getting the pieces to fit together. It is known that increasing front-end analysis reduces the probability of problems later on, but excessive front-end analysis may not pay the anticipated dividends. The key is to accurately estimate early in a program the appropriate level of SE in order to ensure system success within cost and schedule budgets.

Most widely used estimation tools, shown in Table 4, treat SE as a subset of a software or a hardware effort. Since complex systems are not dominated by either hardware or software, SE ought not to be viewed as a subset of hardware or software. Rather, because many functions can be implemented using either hardware or software, SE is becoming the discipline for selecting, specifying and coordinating the various hardware and software designs. Given that role, the correct path is to forecast SE resource needs based on the tasks that systems engineering must perform and not as an arbitrary percentage of another effort. Hence, SE estimation tools must provide for aligning the definition of tasks that SE is expected to do on a given project with the program management's vision of economic and schedule cost, performance, and risk.

Tools that forecast SE resources largely ignore factors that reflect the scope of the SE effort, as insufficient historical data exists from which statistically significant algorithms can be derived. To derive cost-estimating relationships from historical data using regression analysis, one must have considerably more data points than variables; such as a ratio of 5 to 1. It is difficult to obtain actual data on systems engineering costs and on factors that impact those costs. For example, a typical factor may be an aggressive schedule, which will increase the demand for SE resources. The result is a tool set that inadequately characterizes the proposed program and therefore inaccurately forecasts SE resource needs. Moreover, the tools listed in Table 4 use different life cycle stages, complicating things even further. The names of the different life cycle stages and a mapping to each other is provided in Figure 2. The three software models have different life cycle stages than the five hardware models. As a result, only models with similar life cycle phases are mapped to each other.

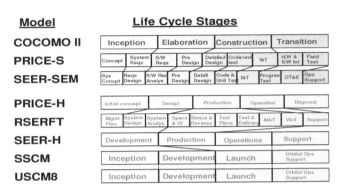

Model	Life Cycle Stages									
COCOMO II	Inception		Elaboration		Construction			Transition		
PRICE-S	Concept	System Reqs	S/W Reqs	Pre Design	Detailed Design	Code/unit test	I&T	H/W & S/W Int	Field Test	
SEER-SEM	Sys Concept	Reqs Design	S/W Req Analys	Pre Design	Detail Design	Code & Unit Test	I&T	Program Test	OT&E	Ops Support
PRICE-H	Initial concept		Design		Production		Operation		Disposal	
RSERFT	Mgmt Plan	System Design	System Analys	Specs & I/F	Status & Reviews	Test Plans	Test & Delivery	AI&T	V&V	Support
SEER-H	Development		Production		Operations			Support		
SSCM	Inception		Development		Launch			Orbital Ops Support		
USCM8	Inception		Development		Launch			Orbital Ops Support		

Figure 2 Model Life Cycle Phases Compared

As the parallels between hardware and software estimation models are drawn and the relationships between these and systems engineering are defined it is easy to identify the pressing need for a model that can estimate systems engineering as an independent function. The fundamental approach for developing a model that meets this demand relates back to the area of software cost estimation from which the theoretical underpinnings of COSYSMO are derived. This area of research is described in the next section.

2.2. COSYSMO Lineage

In order to place COSYSMO in the right context it must be linked to the work that has preceded it. A wealth of models and processes exist in the area of software engineering, from which this work is derived. Particularly the Model-Based System Architecting and Software Engineering (MBASE) framework (Boehm and Port 1999) developed for the purposes of tailoring a software project's balance of discipline and flexibility via risk considerations. As an elaboration of the spiral model (Boehm and Hansen 2001), MBASE provides a framework for projects to use various process, product, property, and success models. *Process models* include the waterfall model, evolutionary development, incremental development, spiral development, rapid application development, and many others. *Product models* include various ways of specifying operational concepts, requirements, architectures, designs, and code, along with their interrelationships. *Property models* include models for cost, schedule, performance, reliability, security, portability, etc., and their tradeoffs. *Success models* include organization and project goals, stakeholder win-win, business-case, or IKIWISI (I'll know it when I see it). COSYSMO is considered a *property model* because it focuses on the

23

effort and cost associated with systems engineering and the tradeoffs between decisions that affect systems engineering. Awareness of COSYSMO's model category can help prevent clashes between other models within or outside of the model category (Boehm and Port 1999). Equally important as COSYSMO's lineage is its link to existing systems engineering estimation methods. It provides valuable context of the state of the practice surrounding it while informing users of the available alternatives.

2.3. Overview of Systems Engineering Estimation Methods

A number of useful systems engineering estimation techniques are currently in use by practitioners. They vary in both maturity and sophistication. Subsequently, some are more easily adaptable to the changing environment and others take more time to develop. The logic behind these approaches is fundamentally different, leaving only their results as measures of merit. It is believed that a hybrid approach that borrows from each method is the best way to capture systems engineering phenomena that a single approach may miss. Six estimation techniques are presented here in order of sophistication.

Heuristics & rules of thumb. Heuristic reasoning has been commonly used by engineers to arrive at quick answers to their questions. Practicing engineers, through education, experience, and examples, accumulate a considerable body of contextual intuition. These experiences evolve into instinct or common sense that are seldom recorded. These can be considered insights, lessons learned, and rules of thumb, among other names, that are brought to bear on certain situations. Ultimately, this knowledge is based on experience and often provides valuable results. Systems engineering cost estimation heuristics and rules of thumb have been developed by researchers and practitioners (Boehm, Abts et al. 2000; Honour 2002; Rechtin 1991). One such rule of thumb, provided by Barry Horowitz, retired CEO of MITRE Corporation, adopts the following logic for estimating systems engineering effort (Horowitz 2004):

If it is a custom developed system (mostly) or an Off-the-Shelf (OTS) integration (mostly)

Then the former gets 6-15% of the total budget for SE, the later gets 15-25% of budget (where selection of OTS products as well as standards is considered SE).

The following additional rules apply:

If the system unprecedented

Then raise the budget from minimum level to 50% more

24

If the system faces an extreme requirement (safety, performance, etc)

Then raise the budget by 25% of minimum

If the system involves a large number of distinct technologies, and therefore a diversity of engineering disciplines and specialties

Then raise the budget by 25% of minimum

If the priority for the system is very high compared to other systems also competing for resources

Then add 50% to the base

Note that the % of SE is larger for OTS, but since the budgets for these projects are much lower, so are the numbers for SE.

Expert opinion. This is the most informal of the approaches because it simply involves querying the experts in a specific domain and taking their subjective opinion as an input. This approach is useful in the absence of empirical data and is very simple. The obvious drawback is that an estimate is only as good as the expert's opinion, which can vary greatly from person to person. However, many years of experience is not a guarantee of future expertise due to new requirements, business processes, and added complexity. Moreover, this technique relies on experts and even the most highly competent experts can be wrong. A common technique for capturing expert opinion is the Delphi (Dalkey 1969) method which was improved and renamed Wideband Delphi (Boehm 1981). This work employs the Wideband Delphi method which is elaborated in section 5.1.

Case studies and analogy. Recognizing that companies do not constantly reinvent the wheel every time a new project comes along, there is an approach that capitalizes on the institutional memory of an organization to develop its estimates. Case studies represent an inductive process, whereby estimators and planners try to learn useful general lessons by extrapolation from specific examples. They examine in detail elaborate studies describing the environmental conditions and constraints that were present during the development of previous projects, the technical and managerial decisions that were made, and the final successes or failures that resulted. They then determine the underlying links between cause and effect that can be applied in other contexts. Ideally, they look for cases describing projects similar to the project for which they will be attempting to develop estimates and apply the rule of analogy that assumes previous performance is an indicator of future performance. The sources of case studies may be either internal or external to the estimator's

25

own organization. Homegrown cases are likely to be more useful for the purposes of estimation because they reflect the specific engineering and business practices likely to be applied to an organization's projects in the future. Well-documented cases studies from other organizations doing similar kinds of work can also prove very useful so long as their differences are identified.

Top Down & Design To Cost. This technique aims for an aggregate estimate for the cost of the project based upon the overall features of the system. Once a total cost is estimated, each subcomponent is assigned a percentage of that cost. The main advantage of this approach is the ability to capture system level effort such as component integration and configuration management. It can also be useful when a certain cost target must be reached regardless of the technical features. The top down approach can often miss the low level nuances that can emerge in large systems. It also lacks detailed breakdown of the subcomponents that make up the system.

Bottom Up & Activity Based. Opposite the top-down approach, bottom-up begins with the lowest level cost component and rolls it up to the highest level for its estimate. The main advantage is that the lower level estimates are typically provided by the people who will be responsible for doing the work. This work is typically represented in the form of a Work Breakdown Structure (WBS), which makes this estimate easily justifiable because of its close relationship to the activities required by the project elements. This can translate to a fairly accurate estimate at the lower level. The disadvantages are that this process is labor intensive and is typically not uniform across entities. In addition, every level folds in another layer of conservative management reserve which can result in an over estimate at the end.

Parametric cost estimation models. This method is the most sophisticated and most difficult to develop. Parametric models generate cost estimates based on mathematical relationships between independent variables (i.e., requirements) and dependent variables (i.e., effort). The inputs characterize the nature of the work to be done, plus the environmental conditions under which the work will be performed and delivered. The definition of the mathematical relationships between the independent and dependent variables is the heart of parametric modeling. These relationships are commonly referred to Cost Estimating Relationships (CERs) and are usually based upon statistical analyses of large amounts of data. Regression models are used to validate the CERs and operationalize them in linear or nonlinear equations. The main advantage of using parametric models is that, once validated, they are fast and easy to use. They do not require a lot of information and can provide fairly

26

accurate estimates. Parametric models can also be tailored to a specific organization's CERs. The major disadvantage of parametric models is that they are difficult and time consuming to develop and require a lot of clean, complete, and uncorrelated data to be properly validated.

As a parametric model, COSYSMO contains its own CERs and is structured in a way to accommodate the current systems engineering standards and processes. Its structure is described in detail in the next section.

3. Model Definition

3.1. COSYSMO Derivation

Since its inception, COSYSMO has gone through three major iterations. This section describes each of these spirals and the properties of the model at those points in time culminating with the final form of the model represented in Equation 6.

3.1.1. Evolution

Spiral #1: Strawman COSYSMO. The first version of COSYSMO contained a list of 16 systems engineering cost drivers. This representation of the model was referred to as the "strawman" version because it provided a skeleton for the model with limited content. The factors identified were ranked by relative importance by a group of experts. Half of the factors were labeled *application factors* and the other half were labeled *team factors*. Each parameter was determined to have a high, medium, or low influence level on systems engineering cost. The most influential application factor was *requirements understanding* and the most influential team factor was *personnel experience.*

Function points and use cases were identified as possible measures of systems engineering functional size. Factors for volatility and reuse were also identified as relevant. At one point the initial list of parameters grew to as many as 24 during one of the brain storming sessions. For reasons related to model parsimony, the number of parameters in the model was eventually reduced from 24 to 18.

Spiral #2: COSYSMO-IP. The second major version of COSYSMO included refined definitions and a revised set of cost drivers. Most importantly, it included measures for functional size that were independent of the software size measures used in COCOMO II. This version had the letters "IP" attached to the end to reflect the emphasis on software "Information Processing" systems as the initial scope. Rooted from interest from industry stakeholders, the focus at the time was to estimate systems engineering effort for software intensive systems. Moreover, this version only covered the early phases of the life cycle: Conceptualize, Develop, and Operational Test & Evaluation. Recognizing that the model had to evolve out of the software intensive arena and on to a broader category of systems, a model evolution plan was developed to characterize the different types of systems that could eventually be estimated with COSYSMO and their corresponding life cycle stages (Boehm, Reifer et al. 2003).

28

The important distinction between size drivers and cost drivers was also clarified. At this stage, a general form for the model was proposed containing three different types of parameters: additive, multiplicative, and exponential.

$$PM = A * (Size)^E * (EM)$$

Equation 1

ADDITIVE   EXPONENTIAL   MULTIPLICATIVE

Where:

    PM = Person Months

    A = calibration factor

    Size = measure(s) of functional size of a system that has an additive effect on
        systems engineering effort

    E = scale factor(s) having an exponential or nonlinear effect on systems
        engineering effort

    EM = effort multipliers that influence systems engineering effort

    The general rationale for whether a factor is additive, exponential, or multiplicative comes from the following criteria (Boehm, Valerdi et al 2005):

1. A factor is additive if it has a local effect on the included entity. For example, adding another source instruction, function point entity, requirement, module, interface, operational scenario, or algorithm to a system has mostly local additive effects. From the additive standpoint, the impact of adding a new item would be inversely proportional to its current size. For example, adding 1 requirement to a system with 10 requirements corresponds to a 10% increase in size while adding the same single requirement to a system with 100 requirements corresponds to a 1% increase in size.

2. A factor is multiplicative if it has a global effect across the overall system. For example, adding another level of service requirement, development site, or incompatible customer has mostly global multiplicative effects. Consider the effect of the factor on the effort associated with the product being developed. If the size of the product is doubled and the proportional effect of that factor is also doubled, then it is a multiplicative factor. For example, introducing a high security requirement to a

29

system with 10 requirements would translate to a 40% increase in effort. Similarly, a high security requirement for a system with 100 requirements would also increase by 40%.

3. A factor that is exponential has both a global effect and an emergent effect for larger systems. If the effect of the factor is more influential as a function of size because of the amount of rework due to architecture, risk resolution, team compatibility, or readiness for SoS integration, then it is treated as an exponential factor.

These statements are pivotal to the hypotheses stated in section 1.3. The next section describes the form of the model and how the hypotheses are tested.

3.1.2. Model Form

Spiral #3: COSYSMO. Substantial insight was obtained from the development of the first two iterations of the model. The current version, referred to simply as COSYSMO, has a broader scope representative of the extensive participation from industrial affiliates and INCOSE. Limiting the boundaries and scope of the model has been one of the most challenging tasks to date, partially because of the features desired by the large number of stakeholders involved in the model development process.

The current operational form of the COSYSMO model is shown in Equation 2. As previously noted, the size drivers and cost drivers were determined via a Delphi exercise by a group of experts in the fields of systems engineering, software engineering, and cost estimation. The definitions for each of the drivers, while not final, attempt to cover those activities that have the greatest impact on estimated systems engineering effort and duration.

$$\text{Equation 2} \qquad PM_{NS} = A \cdot (Size)^{E} \cdot \prod_{i=1}^{n} EM_{i}$$

Where:

PM_{NS} = effort in Person Months (Nominal Schedule)

A = calibration constant derived from historical project data

Size = determined by computing the weighted sum of the four size drivers

E = represents economy/diseconomy of scale; default is 1.0

n = number of cost drivers (14)

$\mathbf{EM}_i$ = effort multiplier for the i_{th} cost driver. Nominal is 1.0. Adjacent multipliers have constant ratios (geometric progression). Within their respective rating scale, the calibrated sensitivity range of a multiplier is the ratio of highest to lowest value.

Each parameter in the equation represents the Cost Estimating Relationships (CERs) that were defined by systems engineering experts. The *Size* factor represents the additive part of the model while the *EM* factor represents the multiplicative part of the model. Specific definitions for these parameters are provided in the following sections.

A detailed derivation of the terms in Equation 2 and motivation for the model is provided here. The dependent variable is the number of systems engineering person months of effort required under the assumption of a nominal schedule, or PM_{NS}. COSYSMO is designed to estimate the number of person months as a function of a system's functional size with considerations of diseconomies of scale. Namely, larger systems will require proportionally more systems engineering effort than smaller systems. That is, larger systems require a larger number of systems engineering person months to complete. The four metrics selected as reliable systems engineering size drivers are: *Number of System Requirements, Number of Major Interfaces, Number of Critical Algorithms,* and *Number of Operational Scenarios.* The weighted sum of these drivers represents a system's functional size from the systems engineering standpoint and is represented in the following CER:

$$\text{Equation 3} \qquad PM_{NS} = \sum_k w_e \Phi_e + w_n \Phi_n + w_d \Phi_d$$

Where:

$\mathbf{k}$ = REQ, INTF, ALG, OPSC

w = weight

$\mathbf{e}$ = easy

$\mathbf{n}$ = nominal

$\mathbf{d}$ = difficult

Φ = driver count

Equation 3 is an operationalization of the four size drivers and includes twelve possible combinations of weights combined with size metrics. Discrete weights for the size drivers, w, can take on the values of "easy", "nominal", and "difficult"; and quantities, Φ, can take on any continuous integer value depending on the number of requirements, interfaces, algorithms, and operational scenarios in the system of interest. All twelve possible combinations may not apply to all systems. This approach of using weighted sums of factors is similar to the software function approach used in other cost models (Albrecht and Gaffney 1983).

The CER shown in Equation 3 is a representation of the relationship between functional size and systems engineering effort. The effect of each size driver on the number of systems engineering person months is determined by its corresponding weight factor. Figure 3 illustrates the relationship between the number of operational scenarios and functional size. This size driver was selected as an example since it was shown to have the highest influence on systems engineering effort.

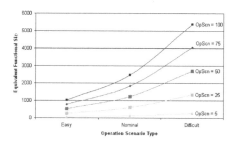

Figure 3 Notional Relationships Between Operational Scenarios
Versus Functional Size

The five curves in Figure 3 are a notional representation of the effects of the weights of the easy, nominal, and difficult operational scenarios on functional size. In addition to functional size there are other people-related emergent properties of systems that arise as larger system-of-systems are created. These properties are similar to the ones previously observed in software development (Banker et al 1994). Different systems engineering efforts may exhibit different levels of productivity which must be represented in COSYSMO. An exponential factor, E, is added to the CER and is represented in Equation 4:

$$\text{Equation 4} \qquad PM_{NS} = \left(\sum_k w_e \Phi_e + w_n \Phi_n + w_d \Phi_d \right)^E$$

32

This factor relates to hypothesis #3. In the case of small projects the exponent, E, could be equal to or less than 1.0. This would represent an economy of scale which is generally very difficult to achieve in large people-intensive projects. Most large projects would exhibit diseconomies of scale and as such would employ a value greater than 1.0 for E. Systems development activities may have different diseconomies of scale because of two main reasons: growth of interpersonal communications overhead and growth of large-system integration overhead. The impact of interpersonal communications has been modeled by researchers in the area of human networks and is believed to be influential in systems engineering. The COCOMO II model includes a diseconomy of scale factor which is approximately 1.1. Other theories suggest that human related diseconomies behave in ways proportional to 2^n, n^2, or n^2-n. A notional example is shown in Figure 4 which includes the actual diseconomies of scale built into COCOMO II and COSYSMO. While the cost models are not as dramatic as theories suggest it must be noted that this parameter only covers human diseconomies. Technical diseconomies are adequately by size and cost drivers.

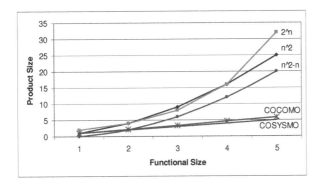

Figure 4 Examples of Diseconomies of Scale

Just as different systems may exhibit various economies of scale, different organizations may exhibit various relationships between systems engineering size and effort. The CER in Equation 5 requires a calibration or adjustment factor that allows for the tuning of COSYSMO to accurately reflect an organization's business line productivity. This factor, A, is included in Equation 5.

33

Equation 5 $\qquad PM_{NS} = A \cdot \left(\sum_k w_e \Phi_e + w_n \Phi_n + w_d \Phi_d \right)^E$

Finally, there is a group of fourteen effort multipliers that have been identified to be significant drivers of systems engineering effort. These are used to adjust the nominal person month effort to reflect the system under development. Each driver is defined by a set of rating levels and corresponding multiplier factors. The nominal level always has an effort multiplier of 1.0, which has no effect on the CER. Off-nominal ratings change the overall estimated effort based on their user-defined values. Equation 6 includes these multiplicative factors, *EM*.

Equation 6 $\qquad PM_{NS} = A \cdot \left(\sum_k w_e \Phi_e + w_n \Phi_n + w_d \Phi_d \right)^E \cdot \prod_{j=1}^{14} EM_j$

Equation 6 is the final COSYSMO CER that was used in the Delphi surveys and historical data collection. Each parameter will be introduced together with its rating scale and counting rules.

3.2. Systems Engineering Size Drivers

The role of the *Size* drivers is to capture the functional size of the system from the systems engineering perspective. They represent a quantifiable characteristic that can be arrived at by objective measures (i.e., physical size). It can be shown that developing a satellite ground station represents a larger systems engineering effort than developing a toaster and in order to differentiate the two, four properties were developed to help quantify the difference. In software cost estimation, some common measures of size include Software Lines of Code (SLOC), Function Points (FP), or Application Points (AP). These sizing approaches contain adjustment factors that give the model the flexibility to estimate software development for different languages running on different platforms. However, when the system involves hardware, software, people, and processes, these measures become insufficient.

Since the focus of this work is systems engineering effort, the size drivers need to apply to software, hardware, and systems containing both. The set of size drivers that affect systems engineering effort were defined to be: *# of Requirements, # of Major Interfaces, # of Critical Algorithms,* and *# of Operational Scenarios.* Originally, three additional size drivers were considered: *# of Modes* (merged with scenarios), *# of Level of Service Requirements,*

34

and *# of design levels* (determined to be multiplicative cost drivers). Of these four, *# of Requirements* has been the most controversial and volatile. This is due in part to the different types of requirements (i.e., functional, operational, environmental) that are used to define systems and their functions, the different levels of requirements decomposition used by organizations, and the varying degree of quality of requirements definition (how well they are written).

The size drivers are quantitative parameters that can be derived from project documentation. Table 5 lists the typical sources that can provide information for each of the four size drivers in COSYSMO.

Table 5 Size Drivers and Corresponding Data Items

Driver Name	Data Item
# of System Requirements	Counted from the system specification
# of Major Interfaces	Counted from interface control document(s)
# of Critical Algorithms	Counted from system spec or mode description docs
# of Operational Scenarios	Counted from test cases or use cases

Early in the system life cycle, these sources may not be available to organizations due to the evolutionary nature of systems. In this case surrogate sources of data must be obtained or derived in order to capture leading indicators related to the four size drivers. Some of these sources may be previous acquisition programs or simulations of future programs.

Each size driver has both continuous and categorical variable attributes. As a continuous variable it can represent a theoretical continuum such as "requirements" or interfaces", which can range from small systems to very large systems of systems; with most cases falling within an expected range. As a categorical variable it can be represented in terms of discrete categories such as "easy" or "difficult" that cannot be measured more precisely. The categorical scales are presented next and the counting rules for determining the values of the continuous variables are provided in the following sections.

Each of the drivers in Table 5 can be adjusted with three factors: volatility, complexity, and reuse. System requirements are frequently volatile and, in a dynamic environment, are expected to increase as the project progresses. This phenomenon, known as scope creep, is commonly quantified by expansion and stability patterns (Hammer et al 1998). Although new requirements are created, deleted, and modified throughout the life cycle of the project, empirical studies suggest that there tends to be an average number of low level requirements that need to be written in order to satisfy the requirements at the previous i.e.

high level. These studies show that the expansion of requirements shows an expected bell curve. Intuitively, it makes sense to implement stable requirements first and hold off on the implementation of the most volatile requirements until late in the development cycle (Firesmith 2004). Any volatility beyond what is normally expected can greatly contribute to an increase in size.

The second factor used to adjust the size drivers of COSYSMO model is the complexity level of the requirements. A typical system may have hundreds, or potentially thousands, of requirements that are decomposed further into requirements pertaining to the next subsystem. Naturally, not all requirements have the same level of complexity. Some may be more complex than others based on how well they are specified, how easily they are traceable to their source, and how much they overlap with other requirements. It has been determined that a simple sum of the total number of requirements is not a reliable indicator of functional size. Instead, the sum of the requirements requires a complexity weight to reflect the corresponding complexity of each requirement. Logically, the more complex a requirement the greater the weight that is assigned to it. It is up to the individual organization to make an assessment of the complexity of the size drivers associated with their systems. Guidance on how to accomplish this for each size driver is provided in the next sections.

Reuse is the third important factor used to adjust the number of requirements. As reuse facilitates the usage of certain components in the system it tends to bring down the efforts involved in the system development. The sum of requirements is adjusted downwards when there are a significant number of reused requirements. This is meant to capture an organization's familiarity with the development, management, and testing of requirements. However, reused requirements are not free from systems engineering effort. There are three components of reuse each of which has a cost: redesign, reimplementation, and retest. Redesign is necessary when the existing functionality may not be exactly suited to the new task. When this is so, the application to be reused will likely require some rework to support new functions, and it may require reverse engineering to understand its current operation. Some design changes may be in order as well. Changing design will also result in reimplementation changes. Even if redesign and reimplementation are not required, retesting is almost always needed to ensure legacy systems operate properly in their new environment. In summary, reuse may adjust the influence of size drivers upwards or downwards depending on the system characteristics. The three adjustment factors are summarized in Table 6.

Table 6 Adjustment Factors for Size Drivers

Adjustment Factor	Influence on Size
Volatility	Increase
Complexity	Increase
Reuse	Increase or decrease

Of the three adjustment factors, complexity was the most useful when characterizing each size driver. Experts found it easier to assign complexity levels to size drivers based on their past experience with systems. To facilitate the assignment of the complexity adjustment factors, a corresponding definition and rating scale was developed for each size driver. The rating scale is divided into three sections: easy, nominal, and difficult; each corresponding to a complexity weight for each of the three levels. Volatility and reuse were left as future add-ons to the model because they were more difficult to obtain expert opinion.

3.2.1. Number of System Requirements

The definition and three adjustment factors alone do not capture all the impact introduced by requirements. Additional work is involved in decomposing requirements so that they may be counted at the appropriate system-of-interest. As part of this work, rules have been developed to help clarify the definition and adjustment factors while providing consistent interpretations of the size drivers for use in cost estimation.

Other data items, or sources, may be available on certain projects depending on the processes used in the organization. For example, system requirements may be counted from the requirements verification matrix or a requirements management tool such as DOORS.

Table 7 Number of System Requirements Definition

Number of System Requirements
This driver represents the number of requirements for the system-of-interest at a specific level of design. The quantity of requirements includes those related to the effort involved in system engineering the system interfaces, system specific algorithms, and operational scenarios. Requirements may be functional, performance, feature, or service-oriented in nature depending on the methodology used for specification. They may also be defined by the customer or contractor. Each requirement may have effort associated with it such as verification and validation, functional decomposition, functional allocation, etc. System requirements can typically be quantified by counting the number of applicable shalls/wills/shoulds/mays in the system or marketing specification. Note: some work is involved in decomposing requirements so that they may be counted at the appropriate system-of-interest.

Table 8 Number of System Requirements Rating Scale

Easy	Nominal	Difficult
- Simple to implement	- Familiar	- Complex to implement or engineer
- Traceable to source	- Can be traced to source with some effort	- Hard to trace to source
- Little requirements overlap	- Some overlap	- High degree of requirements overlap

37

A particular system may have some requirements that could be considered easy because they are straightforward and have been implemented successfully before, some requirements could be nominal because they are moderately complex and require some effort, and some requirements could be difficult because they are very complex and have a high degree of overlap with other requirements.

The challenge with requirements is that they can be specified by either the customer or the contractor. In addition, these organizations often specify system requirements at different levels of decomposition and with different levels of sophistication. Customers may provide high level requirements in the form of system capabilities, objectives, or measures of effectiveness; these are translated into requirements by the contractor and decomposed into different levels depending on the role of the system integrator. The prime contractor could decompose the initial set of requirements and expand them to subcontractors below it as illustrated in Figure 5.

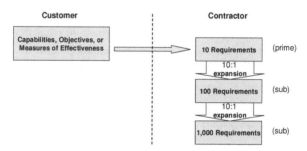

Figure 5 Notional Example of Requirements Translation from Customer to Contractor

For purposes of this example, the expansion ratio from one level of requirement decomposition to the other is assumed to be 10:1. Different systems will exhibit different levels of requirements decomposition depending on the application domain, customer's ability to write good system requirements, and the functional size of the system. The requirements flow framework in Figure 5 provides a starting point for the development of rules to count requirements. These rules were designed to increase the reliability of requirements counting by different organizations on different systems regardless of their application domain. The five rules are as follows:

38

1. **Determine the system of interest.** For an airplane, the system of interest may be the avionics subsystem or the entire airplane depending on the perspective of the organization interested in estimating systems engineering. This key decision needs to be made early on to determine the scope of the COSYSMO estimate and identify the requirements that are applicable for the chosen system.

2. **Decompose system objectives, capabilities, or measures of effectiveness into requirements that can be tested, verified, or designed.** The decomposition of requirements must be performed by the organization using COSYSMO. The level of decomposition of interest for COSYSMO is the level in which the system will be designed and tested; which is equivalent to the TYPE A, System/Segment Specification (MIL-STD 490-A 1985).

3. **Provide a graphical or narrative representation of the system of interest and how it relates to the rest of the system.** This step focuses on the hierarchical relationship between the system elements. This information can help describe the size of the system and its levels of design. It serves as a sanity check for the previous two steps.

4. **Count the number of requirements in the system/marketing specification or the verification test matrix for the level of design in which systems engineering is taking place in the desired system of interest.** The focus of the counted requirements needs to be for systems engineering. Lower level requirements may not be applicable if they have no effect on systems engineering. Requirements may be counted from the Requirements Verification Trace Matrix (RVTM) that is used for testing system requirements. The same rules apply as before: all counted requirements must be at the same design or bid level and lower level requirements must be disregarded if they do not influence systems engineering effort.

5. **Determine the volatility, complexity, and reuse of requirements.** Once the quantity of requirements has been determined, the three adjustment factors can be applied. Currently three complexity factors have been determined: easy, nominal, and difficult. These weights for these factors were determined using expert opinion through the use of a Delphi survey (Valerdi et al 2003). The volatility and reuse factors are optional and depend on the version of COSYSMO implementation being used.

The objective of the five steps is to lead users down a consistent path of similar logic when determining the number of system requirements for the purposes of estimating systems

39

engineering effort in COSYSMO. It has been found that the level of decomposition described in step #2 may be the most volatile step as indicated by the data collected thus far. To alleviate this, a framework of software use case decomposition was adopted (Cockburn 2001). The basic premise behind the framework is that different levels exist for specific system functions. Choosing the appropriate level can provide a focused basis for describing the customer and developer needs. A metaphor is used to describe four levels: *sky level, kite level, sea level,* and *underwater level.* The development of COSYSMO can be used to further illustrate the *sea level* metaphor. The summary level, or *sky level,* represents the highest level that describes either a strategic or system scope.

For example, a *sky level* goal for COSYSMO is to "build a systems engineering cost model." The stakeholders of the model stated this as their basic need that in turn drives a collection of user level goals. A *kite level* goal provides more detailed information as to "how" the *sky level* goal will be satisfied. Continuing the example, it includes the standards that will drive the definition of systems engineering and system life cycle phases. The *sea level* goals represent a user level task that is the target level for counting requirements in COSYSMO. It involves utilizing size and cost drivers, definitions, and counting rules that will enable the accurate estimation of systems engineering effort, also providing more information on how the higher goals at the *kite level* will be satisfied. The *sea level* is also important because it describes the environment in which the model developers interact with the users and stakeholders. A step below is the *underwater level* which is of more concern to the developer. In this example, it involves the selection of implementation and analysis tools required to meet the user goals. The examples are mapped to Cockburn's hierarchy in Figure 6.

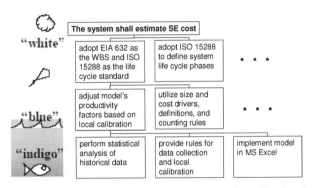

Figure 6 Cockburn's Hierarchy as Related to COSYSMO Use Case Levels
(adapted from Cockburn 2001)

Going down the hierarchy from *sky* to *underwater* provides information on "how" a particular requirement will be satisfied by the system while going up the hierarchy provides information on "why" a particular requirement exists.

3.2.2. Number of System Interfaces

System interfaces are also important drivers of systems engineering effort. The greatest leverage in system architecting is at the interfaces (Rechtin 1991) but this leverage comes at a price. Both the quantity and complexity of interfaces require more systems engineering effort.

Table 9 Number of System Interfaces Definition

Number of System Interfaces
This driver represents the number of shared physical and logical boundaries between system components or functions (internal interfaces) and those external to the system (external interfaces). These interfaces typically can be quantified by counting the number of external and internal system interfaces among ISO/IEC 15288-defined system elements.

Table 10 Number of System Interfaces Rating Scale

Easy	Nominal	Difficult
- Simple message	- Moderate complexity	- Complex protocol(s)
- Uncoupled	- Loosely coupled	- Highly coupled
- Strong consensus	- Moderate consensus	- Low consensus
- Well behaved	- Predictable behavior	- Poorly behaved

41

Similar problems of decomposition exist for this driver because interfaces are defined at multiple levels of the system hierarchy. The target level for counting interfaces involves the following rules:

1. **Focus on technical interfaces only.** Other parameters in the model address organizational interfaces.

2. **Identify the interfaces that involve systems engineering for your system of interest.** Counting interfaces at the integrated circuit level is often too low. Sometimes there may be multiple levels of interfaces connecting higher system elements, lower system elements, and elements at the same level of the system hierarchy.

3. **Determine the number of unique interface types.** If twenty interfaces exist but there are only two types of interfaces, then the relevant number to count is two.

4. **Focus on the logical aspects of the interface.** This provides a better indicator of the complexity of each interface from a systems engineering standpoint. Counting the number of wires in an interface may not be a good indicator. Instead, the protocol used or the timing requirement associated with the interface will be a better indicator.

5. **Determine complexity of each interface.** Bidirectional interfaces count as two interfaces because they require coordination on both ends.

3.2.3. Number of Algorithms

The number and complexity of algorithms is also a useful driver for determining systems engineering size and ultimately effort. Both hardware and software algorithms increase systems engineering activities throughout all phases of the life cycle.

Table 11 Number of System-Specific Algorithms Definition

Number of System-Specific Algorithms
This driver represents the number of newly defined or significantly altered functions that require unique mathematical algorithms to be derived in order to achieve the system performance requirements. As an example, this could include a complex aircraft tracking algorithm like a Kalman Filter being derived using existing experience as the basis for the all aspect search function. Another example could be a brand new discrimination algorithm being derived to identify friend or foe function in space-based applications. The number can be quantified by counting the number of unique algorithms needed to realize the requirements specified in the system specification or mode description document.

Table 12 Number of System-Specific Algorithms Rating Scale

Easy	Nominal	Difficult
- Algebraic	- Straight forward calculus	- Complex constrained optimization; pattern recognition
- Straightforward structure	- Nested structure with decision logic	- Recursive in structure with distributed control
- Simple data	- Relational data	- Noisy, ill-conditioned data
- Timing not an issue	- Timing a constraint	- Dynamic, with timing and uncertainty issues
- Adaptation of library-based solution	- Some modeling involved	- Simulation and modeling involved

Since the influence of algorithms can vary by organization, the process of identifying an algorithm for COSYSMO can also be different. Ultimately there are different sources from which the algorithms can be obtained. For example, during the conceptual stage of a system, there is a limited amount of information available. Only functional block diagrams may be available which can serve as indicators of how many algorithms may exist in the system. As the system design evolves and more uncertainties are resolved, there are more sources available to aid in the estimation of algorithms. Table 13 includes examples of the entities that are available at different stages of the system life cycle and their corresponding attributes that can be used to estimate the number of algorithms. They are listed in typical order of availability; the first entities are typically available during the conceptual stages while the latter ones are available as the system design evolves.

Table 13 Candidate Entities and Attributes for Algorithms

Entities	Attributes
Historical database	# of algorithms
Functional block diagram	# of functions that relate to algorithms
Mode description document	algorithms
Risk analysis	algorithm related risks
System specification	algorithms
Subsystem description documents	algorithms
Configuration baseline	technical notes

The attributes may provide more detailed information about the functions that the algorithms perform. This can aid in determining the complexity of that algorithm, an important step in estimating size for COSYSMO.

Determining the quantity of algorithms in a system can also differ across organizations. System algorithms are unique in the sense that they are highly related to the

43

"# of Requirements" and "# of Interfaces" size drivers. If not explicitly defined up front, the number of algorithms can be derived from a system-level requirement or deduced from the properties of an interface. In terms of systems engineering effort, the existence of an algorithm introduces additional work related to simulation, implementation, test cases, documentation, and support. These activities are illustrated in Figure 7.

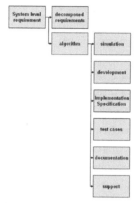

Figure 7 Effort Decomposition Associated With an Algorithm

There exists an entire process in which the general types of algorithms needed are determined, math is developed to implement them, algorithm-related requirements are communicated to other designers (subsystems, hardware, software, etc.) for what data and data quality requirements, and algorithm trade-offs are performed. These activities are within the scope of systems engineering and are covered in COSYSMO.

In some cases, a significant amount of effort associated with systems engineering as related to algorithms will involve reuse which can reduce the complexity of algorithms and in turn the effort associated with their implementation. Conversely, there may be situations where algorithms are unprecedented and loosely defined. From an implementation standpoint, the number of design constraints – such as timing restrictions or processor limitations – may influence the complexity of software algorithms when compared to hardware algorithms. In either case, both types of algorithms should be counted and assigned a level of complexity for input into COSYSMO.

To demonstrate the process of identifying and counting an algorithm an example is provided from the field of signal compression. For purposes of this example it is assumed that a system specification has been developed. From this specification, the following system level requirement is obtained: *All images captured by the sensor shall be compressed in compliance with MPEG-4 coding standard.* This requirement triggers several possible solutions that meet the required standard. A developer may decide to implement the requirement with a well-known algorithm used for compressing visual images: *MPEG-4 Visual Texture Coding (VTC).* As illustrated in Figure 7 this algorithm generates products associated with it which lead to increased systems engineering effort that is estimated by COSYSMO. Other effort generated by the implementation specification, such as software engineering, is not estimated by COSYSMO. Models such as COCOMO II should be used to estimate the software development effort. For purposes of COSYSMO, the MPEG-4 VTC algorithm counts as one distinct algorithm even if it is used multiple times in the same system. Since this is a well known algorithm with predictable behavior it qualifies as an "easy" algorithm.

3.2.4. Number of Operational Scenarios

The fourth and final size driver captures the operational scenarios of a system. The more operational scenarios – and the more complex these scenarios are – the more systems engineering effort will be required.

Table 14 Number of Operational Scenarios Definition

Number of Operational Scenarios
This driver represents the number of operational scenarios that a system must satisfy. Such scenarios include both the nominal stimulus-response thread plus all of the off-nominal threads resulting from bad or missing data, unavailable processes, network connections, or other exception-handling cases. The number of scenarios can typically be quantified by counting the number of system test thread packages or unique end-to-end tests used to validate the system functionality and performance or by counting the number of use cases, including off-nominal extensions, developed as part of the operational architecture.

Table 15 Number of Operational Scenarios Rating Scale

Easy	Nominal	Difficult
- Well defined	- Loosely defined	- Ill defined
- Loosely coupled	- Moderately coupled	- Tightly coupled or many dependencies/conflicting requirements
- Timelines not an issue	- Timelines a constraint	- Tight timelines through scenario network
- Few, simple off-nominal threads	- Moderate number or complexity of off-nominal threads	- Many or very complex off-nominal threads

In a similar way requirements were defined at *sea level*, operational scenarios must also be identified at a level that is of interest to systems engineering. An example of a typical target level for operational scenarios is shown in Figure 8.

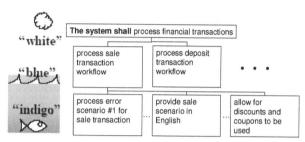

Figure 8 Operational Scenario Example
(adapted from Cockburn 2001)

Hypothesis #1 proposes that the weighted sum of these drivers is a good predictor of size and as size increases the amount of systems engineering effort also increases. Given that the calculation for size is a composite measure of the four size drivers it is evident that a system can be considered large a number of different ways. For example, a system with multiple independent interfaces and very few requirements can be similar in size to a system with few independent interfaces and many requirements.

The use of the words "multiple", "very few", "few", and "many" add a dimension of complexity as the subjectivities in the attributes is quantified. Great care must be taken to use consistent interpretations of these words on systems being estimated. The cost drivers in the model, somewhat subjective in nature, attempt to capture the most important system and development environment characteristics that drive systems engineering cost.

3.3. Systems Engineering Cost Drivers

The cost drivers in the model represent the multiplicative part of the model introduced in Section 3.1. These drivers are also referred to as effort multipliers since they affect the entire systems engineering effort calculation in a multiplicative manner. Assigning ratings for these drivers is not as straight forward as the size drivers mentioned previously. The difference is that most of the cost drivers are qualitative in nature and require subjective assessment in order to be rated. Table 16 shows the data items or information needed in order to assess the cost drivers.

Table 16 Cost Drivers and Corresponding Data Items

Driver Name	Data Item
Requirements understanding	Subjective assessment of the system requirements
Architecture understanding	Subjective assessment of the system architecture
Level of service requirements	Subjective difficulty of satisfying the key performance parameters
Migration complexity	Influence of legacy system (if applicable)
Technology risk	Maturity, readiness, and obsolescence of technology
Documentation to match life cycle needs	Breadth and depth of required documentation
# and Diversity of installations/platforms	Sites, installations, operating environment, and diverse platforms
# of Recursive levels in the design	Number of applicable levels of the Work Breakdown Structure
Stakeholder team cohesion	Subjective assessment of all stakeholders
Personnel/team capability	Subjective assessment of the team's intellectual capability
Personnel experience/continuity	Subjective assessment of staff consistency
Process capability	CMMI level or equivalent rating
Multisite coordination	Location of stakeholders and coordination barriers
Tool support	Subjective assessment of SE tools

In the COCOMO II model, an ensemble of similar drivers is used to modify the amount of effort to reflect product, platform, personnel, and project factors that have been shown to influence cost and schedule for software projects. It has been determined that these themes were not appropriate for systems engineering. New themes have been developed that aggregate the unique phenomena observed in systems engineering. These themes are:

- Understanding. Drivers that capture the level of comprehension and familiarity of the systems engineering team.
- Complexity. Drivers that capture the difficulty, risk, and program-related factors that can influence systems engineering effort.
- Operations. Drivers that capture the planning involved with the implementation from the systems engineering standpoint.
- People. Drivers that capture the capability of the systems engineering team.

- Environment. Drivers that capture the level of sophistication under which systems engineering is being performed.

The criteria for assigning cost drivers to these themes included driver polarity and correlation. Drivers that were associated with productivity savings such as "understanding" or "capability" drivers were grouped together while those associated with productivity penalties such as "complexity" were grouped together. Moderately correlated parameters were also combined based on the results from the correlation matrix in Appendix F.

Hypothesis #2 proposes that clusters of these drivers are accurate predictors of systems engineering effort. Different permutations of these drivers were compared and the best set, shown in Figure 9, was selected based on the aforementioned criterion.

UNDERSTANDING FACTORS
- Requirements understanding
- Architecture understanding
- Stakeholder team cohesion
- Personnel experience/continuity

COMPLEXITY FACTORS
- Level of service requirements
- Technology Risk
- # of Recursive Levels in the Design
- Documentation Match to Life Cycle Needs

OPERATIONS FACTORS
- # and Diversity of Installations/Platforms
- Migration complexity

PEOPLE FACTORS
- Personnel/team capability
- Process capability

ENVIRONMENT FACTORS
- Multisite coordination
- Tool support

Figure 9 Cost Driver Clustering

In addition to a description, each driver was assigned a rating scale that described different attributes that could be used to rate the degree of impact on systems engineering effort. Rating levels included: Very Low, Low, Nominal, High, Very High, and Extra High. The Nominal level represents zero impact on productivity and is therefore assigned a multiplier of 1.0. Levels above and below nominal are assigned multipliers above or below 1.0 to reflect their impact on systems engineering effort. The increase or decrease of multipliers along the rating scale will depend on the polarity of each driver. For example, the requirements understanding is defined in such a way that *Very Low* understanding will have a productivity penalty on systems engineering. As a result, it will have a multiplier of greater than 1.0, such as 1.87, to reflect an 87% productivity penalty. The multipliers for the rating scaled are provided in section 5.1.

3.3.1. Understanding Factors

The first cost driver theme deals with the systems engineering team's comprehension of and familiarity with the system of interest. Higher ratings for these drivers represent a productivity savings. There are four understanding factors, the most influential being *Requirements Understanding*.

Table 17 Requirements Understanding Definition

Requirements understanding
This cost driver rates the level of understanding of the system requirements by all stakeholders including systems, software, hardware, customers, team members, users, etc. Primary sources of added systems engineering effort are unprecedented systems, unfamiliar domains, or systems whose requirements are emergent with use.

Table 18 Requirements Understanding Rating Scale

Very low	Low	Nominal	High	Very High
Poor: emergent requirements or unprecedented system	Minimal: many undefined areas	Reasonable: some undefined areas	Strong: few undefined areas	Full understanding of requirements, familiar system

Counting the number of requirements and rating their complexities is addressed by the size driver. But the overall degree of understanding of these requirements – by all the stakeholders – has a multiplicative effect on the total amount of effort needed for systems engineering.

Table 19 Architecture Understanding Definition

Architecture understanding
This cost driver rates the relative difficulty of determining and managing the system architecture in terms of platforms, standards, components (COTS/GOTS/NDI/new), connectors (protocols), and constraints. This includes tasks like systems analysis, tradeoff analysis, modeling, simulation, case studies, etc.

Table 20 Architecture Understanding Rating Scale

Very low	Low	Nominal	High	Very High
Poor understanding of architecture and COTS, unprecedented system	Minimal understanding of architecture and COTS, many unfamilar areas	Reasonable understanding of architecture and COTS, some unfamiliar areas	Strong understanding of architecture and COTS, few unfamiliar areas	Full understanding of architecture, familiar system and COTS
>6 level WBS	5-6 level WBS	3-4 level WBS	2 level WBS	

Understanding the architecture is also an important part of being able to design the system (Rechtin 1991). The understanding of the system architecture is different than the understanding of requirements and therefore warrants its own driver. Besides unprecedentedness and domain unfamiliarity, primary sources of added systems engineering

49

effort are new technologies, complex COTS products and choices, and depth of the product hierarchy or Work Breakdown Structure (WBS).

Table 21 Stakeholder Team Cohesion Definition

Stakeholder team cohesion
Represents a multi-attribute parameter which includes leadership, shared vision, diversity of stakeholders, approval cycles, group dynamics, Integrated Product Team framework, team dynamics, trust, and amount of change in responsibilities. It further represents the heterogeneity in stakeholder community of the end users, customers, implementers, and development team.

Table 22 Stakeholder Team Cohesion Rating Scale

	Very Low	Low	Nominal	High	Very High
Culture	Stakeholders with diverse expertise, task nature, language, culture, infrastructure Highly heterogeneous stakeholder communities	Heterogeneous stakeholder community Some similarities in language and culture	Shared project culture	Strong team cohesion and project culture Multiple similarities in language and expertise	Virtually homogeneous stakeholder communities Institutionaliz ed project culture
Compatibili ty	Highly conflicting organizational objectives	Converging organizational objectives	Compatible organizational objectives	Clear roles & responsibilities	Strong mutual advantage to collaboration
Familiarity and trust	Lack of trust	Willing to collaborate, little experience	Some familiarity and trust	Extensive successful collaboration	Very high level of familiarity and trust

The mutual culture, compatibility, familiarity, and trust of the stakeholders involved in the development of the system are key project factors that have significant importance in the systems engineering domain. The group's ability to work together is a factor that has been highlighted as being important for software system development (Brooks 1995), and analyzed as a significant software cost driver in COCOMO II (Boehm et al 2000).

Table 23 Personnel Experience/Continuity Definition

Personnel experience/continuity
The applicability and consistency of the staff at the initial stage of the project with respect to the domain, customer, user, technology, tools, etc.

Table 24 Personnel Experience/Continuity Rating Scale

	Very low	Low	Nominal	High	Very High
Experience	Less than 2 months	1 year continuous experience, other technical experience in similar job	3 years of continuous experience	5 years of continuous experience	10 years of continuous experience
Annual Turnover	48%	24%	12%	6%	3%

The team experience rating measures the systems engineers' experience relevant to the system of interest and its context. It should be noted that often times many years of experience does not translate to competency in a certain area. Experience is rated as of the beginning of the project and is expected to increase as the project goes on, unless adversely affected by personnel turnover.

3.3.2. Complexity Factors

Complexity factors account for variation in effort required to develop systems caused by the characteristics of the system under development. A system that has multiple "ilities", immature technology, a complex design, and excessive documentation will require more effort to complete. There are four complexity factors, the most influential being *Level of Service Requirements.*

Table 25 Level of Service Requirements Definitions

Level of service requirements
This cost driver rates the difficulty and criticality of satisfying the ensemble of level of service requirements, such as security, safety, response time, interoperability, maintainability, Key Performance Parameters (KPPs), the "ilities", etc.

Table 26 Level of Service Requirements Rating Scale

	Very low	Low	Nominal	High	Very High
Difficulty	Simple; single dominant KPP	Low, some coupling among KPPs	Moderately complex, coupled KPPs	Difficult, coupled KPPs	Very complex, tightly coupled KPPs
Criticality	Slight inconvenience	Easily recoverable losses	Some loss	High financial loss	Risk to human life

The level of service requirements, "ilities", or Key Performance Parameters, provide an indication of the complexity of the systems engineering effort required to meet all of the stakeholder requirements. The "ilities" may include: reliability, usability, performance, affordability, maintainability, and so forth. The "ilities" are imperatives of the external world as expressed at the boundaries with the internal world of the system (Rechtin 1991).

Reliability, usability, and performance are the imperative of the user, affordability that of the client, and maintainability that of the operator. This driver has two different viewpoints, *difficulty* and *criticality*, that help represent the two dimensions associated with this driver. Ratings for these viewpoints are often not the same. In cases where systems may have a high degree of difficulty in meeting a response time requirement, there is an equally severe level of criticality associated with not meeting it.

Table 27 Technology Risk Definition

Technology Risk
The maturity, readiness, and obsolescence of the technology being implemented. Immature or obsolescent technology will require more Systems Engineering effort.

Table 28 Technology Risk Rating Scale

	Very Low	Low	Nominal	High	Very High
Lack of Maturity	Technology proven and widely used throughout industry	Proven through actual use and ready for widespread adoption	Proven on pilot projects and ready to roll-out for production jobs	Ready for pilot use	Still in the laboratory
Lack of Readiness	Mission proven (TRL 9)	Concept qualified (TRL 8)	Concept has been demonstrated (TRL 7)	Proof of concept validated (TRL 5 & 6)	Concept defined (TRL 3 & 4)
Obsolescence			- Technology is the state-of-the-practice - Emerging technology could compete in future	- Technology is stale - New and better technology is ready for pilot use	- Technology is outdated and use should be avoided in new systems - Spare parts supply is scarce

Another attribute of the project may be the risk being employed by adopting a certain technology. Some work has been done to show the negative effects of technologies over a long time horizon (Valerdi and Kohl 2004) and frameworks have been developed to show how products with short life cycles can affect the overall project risk (Smith 2004). The maturity of the technology or lack thereof, has a significant effect on the amount of systems engineering effort required on a project. In addition, too mature or obsolete technology can increase the necessary amount of systems engineering effort.

Table 29 Number of Recursive Levels in the Design Definition

# of recursive levels in the design
The number of levels of design related to the system-of-interest (as defined by ISO/IEC 15288) and the amount of required SE effort for each level.

Table 30 Number of Recursive Levels in the Design Rating Scale

	Very Low	Low	Nominal	High	Very High
Number of levels	1	2	3-5	6-7	>7
Required SE effort	Focused on single product	Some vertical and horizontal coordination	More complex interdependencies coordination, and tradeoff analysis	Very complex interdependencies coordination, and tradeoff analysis	Extremely complex interdependencies coordination, and tradeoff analysis

Larger and more complex systems require more systems engineering effort because of the growing number of complexities involved in horizontal and vertical requirements negotiation, tradeoff analyses, architecting, interface definition, scheduled coordination, risk management, and integration. In principle, the systems engineering effort required for a particular system depends solely on the system-of-interest included below it in the design hierarchy, as defined in ISO/IEC 15288. However, a 2-level system within a 4-level system of systems may have more coordination and integration concerns than a standalone 2-level system.

Table 31 Documentation Match to Life Cycle Needs Definition

Documentation match to life cycle needs
The formality and detail of documentation required to be formally delivered based on the life cycle needs of the system.

Table 32 Documentation Match to Life Cycle Needs Rating Scale

	Very low	Low	Nominal	High	Very High
Formality	General goals, stories	Broad guidance, flexibility is allowed	Risk-driven degree of formality	Partially streamlined process, largely standards-driven	Rigorous, follows strict standards and requirements
Detail	Minimal or no specified documentation and review requirements relative to life cycle needs	Relaxed documentation and review requirements relative to life cycle needs	Risk-driven degree of formality, amount of documentation and reviews in sync and consistent with life cycle needs of the system	High amounts of documentation, more rigorous relative to life cycle needs, some revisions required	Extensive documentation and review requirements relative to life cycle needs, multiple revisions required

Some system products require large amounts of documentation. The importance and enforcement of this requirement is often related to the type of system being developed and the nature of the system's users, operators, and maintainers. Less documentation is needed when the users and operators are expert and stable, and the developers become the

53

maintainers. The importance of this factor, then, is based on the match or mismatch of documentation requirements to the life cycle needs of the system. Attempting to save costs via very low or low documentation levels will generally incur extra costs during the maintenance portion of the life cycle. Poor or missing documentation can also cause additional problems in other stages of the life cycle. This driver includes two dimensions, *formality* and *detail*, to represent the different aspects of documentation that need to be considered. A nominal rating involves documentation consistent with life cycle needs. Its degree of formality and detail are risk-driven; if it is low risk not to include something, it is not included.

3.3.3. Operations Factors

The operations factors refer to the hardware and software environments that a system will operate within. Depending on the system of interest, the platform might be an aircraft carrier; an aircraft; an airborne missile; a navigation, guidance, and control system; or a level of the computer system's software infrastructure. The existence of legacy issues may also impact the amount of systems engineering effort required to incorporate the new system with existing technologies and cultures.

Table 33 Number and Diversity of Installations/Platforms Definition

and diversity of installations/platforms
The number of different platforms that the system will be hosted and installed on. The complexity in the operating environment (space, sea, land, fixed, mobile, portable, information assurance/security, constraints on size weight, and power). For example, in a wireless network it could be the number of unique installation sites and the number of and types of fixed clients, mobile clients, and servers. Number of platforms being implemented should be added to the number being phased out (dual count).

54

Table 34 Number and Diversity of Installations/Platforms Rating Scale

	Nominal	High	Very High	Extra High
Sites/ installations	Single installation site or configuration	2-3 sites or diverse installation configurations	4-5 sites or diverse installation configurations	>6 sites or diverse installation configurations
Operating environment	Existing facility meets all known environmental operating requirements	Moderate environmental constraints; controlled environment (i.e., A/C, electrical)	Ruggedized mobile land-based requirements; some information security requirements. Coordination between 1 or 2 regulatory or cross functional agencies required.	Harsh environment (space, sea airborne) sensitive information security requirements. Coordination between 3 or more regulatory or cross functional agencies required.
Platforms	<3 types of platforms being installed and/or being phased out/replaced	4-7 types of platforms being installed and/or being phased out/replaced	8-10 types of platforms being installed and/or being phased out/replaced	>10 types of platforms being installed and/or being phased out/replaced
	Homogeneous platforms	Compatible platforms	Heterogeneous, but compatible platforms	Heterogeneous, incompatible platforms
	Typically networked using a single industry standard protocol	Typically networked using a single industry standard protocol and multiple operating systems	Typically networked using a mix of industry standard protocols and proprietary protocols; single operating systems	Typically networked using a mix of industry standard protocols and proprietary protocols; multiple operating systems

A particular system may have significant platform considerations if it has to address many installations or configurations, it is required to operate in unprecedented environments, or it has to accommodate many heterogeneous platforms. These three viewpoints are represented in the driver rating scale.

Table 35 Migration Complexity Definition

Migration complexity
This cost driver rates the extent to which the legacy system affects the migration complexity, if any. Legacy system components, databases, workflows, environments, etc., may affect the new system implementation due to new technology introductions, planned upgrades, increased performance, business process reengineering, etc.

Table 36 Migration Complexity Rating Scale

	Nominal	High	Very High	Extra High
Legacy contractor	Self; legacy system is well documented. Original team largely available	Self; original development team not available; most documentation available	Different contractor; limited documentation	Original contractor out of business; no documentation available
Effect of legacy system on new system	Everything is new; legacy system is completely replaced or non-existent	Migration is restricted to integration only	Migration is related to integration and development	Migration is related to integration, development, architecture and design

55

The presence of a legacy system can introduce multiple aspects of effort that are related to the contractor of the original system, the number of sites or installations, the operating environment, the percent of legacy components that are being affected, and the cutover requirements that affect the new system. If the project has no significant legacy system concerns, a Nominal rating is given.

3.3.4. People Factors

People factors have a strong influence in determining the amount of effort required to develop a system. These factors are for rating the systems engineering team's vs. individual's capability and experience and for rating the project's process capability.

Table 37 Personnel/Team Capability Definition

Personnel/team capability
Composite intellectual capability of a team of Systems Engineers (compared to the national pool of SEs) to analyze complex problems and synthesize solutions.

Table 38 Personnel/Team Capability Rating Scale

Very Low	Low	Nominal	High	Very High
15th percentile	35th percentile	55th percentile	75th percentile	90th percentile

The team capability driver combines the intellectual horsepower of the team members, how much of the workday horsepower is focused on the problems, and the extent to which the horsepower is pulling in compatible directions. It is measured with respect to an assumed national or global distribution of team capabilities.

Table 39 Process Capability Definition

Process capability
The consistency and effectiveness of the project team at performing SE processes. This may be based on assessment ratings from a published process model (e.g., CMMI, EIA-731, SE-CMM, ISO/IEC15504). It can alternatively be based on project team behavioral characteristics, if no assessment has been

Table 40 Process Capability Rating Scale

	Very Low	Low	Nominal	High	Very High	Extra High
Assessment Rating	Level 0 (if continuous model)	Level 1	Level 2	Level 3	Level 4	Level 5
Project Team Behavioral Characteristics	Ad Hoc approach to process performance	Performed SE process, activities driven only by immediate contractual or customer requirements, SE focus limited	Managed SE process, activities driven by customer and stakeholder needs in a suitable manner, SE focus is requirements through design, project-centric approach – not driven by organizational processes	Defined SE process, activities driven by benefit to project, SE focus is through operation, process approach driven by organizational processes tailored for the project	Quantitatively Managed SE process, activities driven by SE benefit, SE focus on all phases of the life cycle	Optimizing SE process, continuous improvement, driven by system engineering and organizational benefit, SE focus is product life cycle & strategic applications
SEMP Sophistication	Management judgment is used	SEMP is used in an ad-hoc manner only on portions of the project that require it	Project uses a SEMP with some customization	Highly customized SEMP exists and is used throughout the organization	The SEMP is thorough and consistently used; organizational rewards are in place for those that improve it	Organization develop best practices for SEMP; all aspects of the project are included in the SEMP; organizational rewards exist for those that improve it

The procedure for determining a project's systems engineering process capability is organized around the Software Engineering Institute's Capability Maturity Model Integration® (CMMI). The time period for rating process capability is the time the project starts, and should be a reflection of the project only – not of the total organization's maturity level. There are two methods of rating process capability. The first captures the CMMI® continuous model (CMMI 2002; Shrum 2000) and the systems engineering maturity model EIA 731 (ANSI/EIA 2002). The second combines the degrees of mastery of each process area as done for the software CMM in COCOMO II (Clark 1997; Boehm et al 2000). Only the first method is employed here. The project team behavioral characteristics are somewhat analogous to the CMMI levels and can be used by organizations that do not use CMMI ratings. The final viewpoint captures levels of sophistication of the Systems Engineering Management Plan. The higher the sophistication of this document, the higher the systems engineering effort savings because the level of planning associated with a SEMP is indicative of well-managed systems engineering processes.

3.3.5. Environment Factors

The environment factors capture the sophistication of the systems engineering environment in a project. Coordination and support are the two drivers that make up this theme.

Table 41 Multisite Coordination Definition

| **Multisite coordination** |
| Location of stakeholders, team members, resources, corporate collaboration barriers. |

Table 42 Multisite Coordination Rating Scale

	Very Low	Low	Nominal	High	Very High	Extra High
Collocation	International, severe time zone impact	Multi-city and multi-national, considerable time zone impact	Multi-city or multi-company, some time zone effects	Same city or metro area	Same building or complex, some co-located stakeholders or onsite representation	Fully co-located stakeholders
Communications	Some phone, mail	Individual phone, FAX	Narrowband e-mail	Wideband electronic communication	Wideband electronic communication, occasional video conference	Interactive multimedia
Corporate collaboration barriers	Severe export and security restrictions	Mild export and security restrictions	Some contractual & Intellectual property constraints	Some collaborative tools & processes in place to facilitate or overcome, mitigate barriers	Widely used and accepted collaborative tools & processes in place to facilitate or overcome, mitigate barriers	Virtual team environment fully supported by interactive, collaborative tools environment

Given the increasing frequency of multisite developments, and indications that these developments require a significant amount of coordination, it is important to account for their impact on systems engineering. Determining the rating for this driver involves assessing three factors: collocation, communications, and corporate collaboration barriers.

Table 43 Tool Support Definition

| **Tool support** |
| Coverage, integration, and maturity of the tools in the Systems Engineering environment. |

Table 44 Tool Support Rating Scale

Very low	Low	Nominal	High	Very High
No SE tools	Simple SE tools, little integration	Basic SE tools moderately integrated throughout the systems engineering process	Strong, mature SE tools, moderately integrated with other disciplines	Strong, mature proactive use of SE tools integrated with process, model-based SE and management systems

Systems engineering includes the use of tools that perform simulation, modeling, optimization, data analysis, requirements traceability, design representation, configuration management, document extraction, etc. The role of tools and integrated support environments have been shown to be influential in software system development (Baik 2000). Similarly, the use of extensive, well-integrated, mature tool support can improve systems engineering productivity. The effort to tailor such tools to a given project is included in COSYSMO estimates, but the effort to develop major new project-specific tools is not. This effort can be adequately covered in COCOMO II or other software cost estimation models.

Equally important to the model size and cost drivers is the process that was used to arrive at the definitions and rating scales. The next section provides insight on the methodology used to develop the current form of COSYSMO and some limitations that are associated with the model.

4. Methodology

4.1. Research Design & Data Collection

The previous sections outlined the need for COSYSMO and outlined the details of the individual drivers that affect systems engineering. The next steps in developing a useful model involve (1) the identification of the appropriate research designs and approaches; and (2) the correct application of these methods to the research question.

Historically, research in the behavioral sciences goes back to the emergence of psychology as a scientific discipline in the 1800's. As psychology matured, it evolved into a collection of methodologies which included scientific inquiry, measurement, and data analysis (Freud 1924). Branches of psychology emerged and provided different analytical techniques which incorporated statistical models and experimental techniques (Berne 1964). Other behavior-oriented fields such as sociology have contributed much in the sense of formal research methods and research design (Babbie 2004). The field of education has provided frameworks for categorizing different types of research designs, methods, and strategies (Isaac and Michael 1997). In this light, the research approach adopted for this work is a combination of field research and quasi-experimental research. The nature of the research question – how to estimate systems engineering – played the major role in determining the selection of these approaches.

Research Design. The purpose of field research design is to study the background, current status, and environmental interactions of a given social unit. The social units of interest are organizations that develop large-scale technology enabled systems and the systems engineers that work on them. The expert data that has been collected through the Delphi survey attempts to capture project phenomena to help understand the role of systems engineering in an organization. The strengths of this method are that it provides:

- an in-depth investigation of systems engineering organizations
- useful anecdotes or examples to illustrate more generalized statistical findings
- observations of real world phenomena and opportunities to incorporate them into theory

The purpose of quasi-experimental research design is to approximate the conditions of the true experiment in a setting that does not allow control or manipulation of all relevant variables. The multiple factors that affect the conditions can compromise the validity of the design. Since the desired social unit – the systems engineering organization – is influenced by multiple outside forces such as corporate culture, customer pressures, financial priorities, employment challenges, and technical obstacles, it is nearly impossible to control all of the conditions. However, this method is useful because it allows for:

- investigation of cause-and-effect relationships
- variance of different type of projects operating under different conditions
- opportunity to test specific hypotheses

Combining the strengths of field research and quasi-experimental research provides significant benefits because they use different perspectives in the data collection process. Having the right frame of mind while defining the hypotheses and then testing them is also extremely important. During the development of COSYSMO, two different research approaches were adopted: interpretivist and positivist. These two techniques provide fundamentally different approaches at gathering data and validating it, but using them at different stages of the research process can enable a more complete study (Klein and Myers 1999).

Research Approach. The interpretivist approach focuses on the complexity of human sense making as the situation emerges. It enables researchers to learn as much as possible about the phenomena being studied and arrive at qualitative conclusions as to the most important factors. The interpretivist approach was used when developing the size and cost driver definitions with USC corporate affiliates. Through a series of interviews, surveys, and working group meetings the identification and definition of the most significant size and cost drivers was accomplished. The three criteria for the interpretivist approach are:

- Credibility – establishing a match between the constructed realities of systems engineering and the respondents or stakeholders
- Transferability – presenting sufficiently detailed systems engineering cost drivers as to enable judgment that these findings can be transferred to other contexts
- Confirmability – Ensuring that the size and cost drivers are grounded in systems engineering theory and not just a result of the researcher's imagination

Once the systems engineering drivers were defined, there was a shift in the research strategy to a positivist approach. The positivist approach focuses on making formal propositions, quantifiable measures of variables, hypothesis testing, and the drawing of inferences about a phenomenon from a representative sample to a stated population. The criteria associated with this approach are:

- Construct validity – establishing the right operational measures for systems engineering size and cost
- Internal validity – establishing causal relationships between the drivers and systems engineering effort
- External validity – establishing the domain to which the systems engineering drivers can be generalized
- Reliability – ensuring that the relationships between the size and cost drivers can be repeated with the same results

These are discussed in more detail in Chapter 5. The shift from interpretivist to positivist is analogous to a qualitative to quantitative shift in research. While the beginning of the model building process required an open mind about the relationships between the size and cost drivers to effort, the latter part of the process involved testing the hypotheses previously defined and determining their suitability to the research questions. Using both of these approaches increases the chances of obtaining interesting findings.

The two research approaches and two strategies were applied to the same methodology used to develop the COCOMO model. This proven model development process is outlined in the book *Software Cost Estimation With COCOMO II* (Boehm, Abts et al. 2000) and illustrated in Figure 10. The methodology has also been used to create the COCOTS (Abts, Boehm et al. 2001) and the COQUALMO (Baik, Boehm et al. 2002) models, among others.

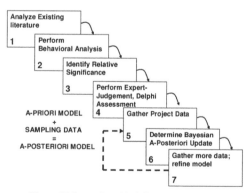

Figure 10 Seven Step Modeling Methodology

The methodology involves (1) analyzing existing literature for factors that affect systems engineering, (2) performing behavioral analyses to determine how the project is behaviorally different with respect to activities performed if the parameter has a higher versus lower rating, (3) identifying the relative significance of the factors on the quantities being estimated, (4) performing the expert-judgment Delphi assessment of the model parameters, (5) gathering historical project data and determining statistical significance of the various parameters, (6) determining the Bayesian posterior set of model parameters, and (7) gathering more data to refine the model based on the experiences.

Steps 2, 4, 5, and 7 employ the field research design because they involve interviews, surveys, and refinement of hypotheses. Steps 3, 6, and 7 employ the quasi-experimental design because they involve the verification of the hypotheses. The interpretivist approach is used in steps 1 and 2 as the research question is approached with an open mind and the model is defined. The positivist approach is used in steps 3, 4, 5, 6, and 7 because they involve the validation of the hypotheses. The use of these designs and approaches, and their relationship to the seven step methodology is summarized in Table 45.

Table 45 Research Designs and Approaches Used
in the 7-step Modeling Methodology

	Step1: Analyze existing literature	Step 2: Perform behavioral analysis	Step 3: Identify relative significance	Step 4: Perform expert judgment; Delphi assessment	Step 5: Gather project data	Step 6: Determine Bayesian A-Posteriori update	Step 7: Gather more data; refine model
Field research design		•		•	•		•
Quasi-experimental research design			•			•	•
Interpretivist approach	•	•					
Positivist approach			•	•	•	•	•

USC CSSE Corporate Affiliate Program. Leveraging off the strong relationships with industry, a working group of 15 core members was assembled to begin the development of the initial version of COSYSMO and identify possible sources of data to use for calibration of the model. Since that time over a dozen more CSSE affiliate organizations have joined the working group and have participated in various working group meetings to refine the model. A COSYSMO e-mail distribution list has been created which contains over 100 subscribers. This distribution list serves as the main communication channel for information pertaining to COSYSMO and upcoming working group meetings. The diverse

63

experience of the working group members includes but is not limited to space systems hardware, information technology, radars, satellite ground stations, and military aircraft. This broad scope helps ensure that the model is robust enough to address multiple areas.

The typical involvement of affiliate companies is twofold. First, each company provides a group of systems engineering experts to rate the model drivers through the use of a wideband Delphi survey. This exercise allows for expert judgement to be captured and included in the model. An additional source of expertise has been the members of INCOSE who have provided extensive valuable feedback that has greatly improved the model. Second, the Affiliate companies provide historical project data for the COSYSMO calibration to validate the model parameters. This ensures that the Cost Estimating Relationships (CERs) in the model are appropriately weighed according to the data received from completed projects.

Need for industry data to calibrate COSYSMO. Industrial participation in the development of COSYSMO is key to the usefulness and relevance of the model. Each driver has a corresponding item that can provide the necessary data for the calibration. The initial industry calibration is essential to understanding the model's robustness, establishing initial relationships between parameters and outcomes, and determining the validity of drivers. However, each organization using COSYSMO will need to perform a local calibration. Through the industry calibration, the working group can establish the values for various scale factors for each driver. This might not be possible or feasible from a local calibration due to the size of the calibration data set and the narrow scope of a single organization's project database. The industry data can also identify elements or features of the model that need refinement. Obtaining data from multiple sources may also identify new drivers that need to be included in future revisions of the model.

An additional important reason for an industry-level calibration is the acceptance of the model for cost estimation by the Defense Contract Audit Agency (DCAA). Even though each organization needs to prove the local calibration matches the local organization's productivity and trends, the industry calibration shows DCAA the model meets the expectations and standards of the Systems Engineering industry. Ensuring that COSYSMO is compatible with these standards plays an important role in its widespread acceptance and generalizability.

Data Collection. The collection of data itself, steps 4, 5, and 7, can be divided into two unique efforts: one focusing on expert data and the other on historical project data. The process used for collecting expert data, the Delphi technique (Dalkey 1969) is performed in Step 4. Developed at The RAND Corporation in the late 1940s, it serves as a way of making predictions about future events - thus its name, recalling the divinations of the Greek oracle of antiquity, located on the southern flank of Mt. Parnassus at Delphi (Ahern, Clouse et al. 2004). More recently, the technique has been used as a means of guiding a group of informed individuals to a consensus of opinion on some issue.

Participants are asked to make an assessment regarding the ratings of size and cost drivers, individually in a preliminary round, without consulting the other participants in the exercise. The first round results are then collected, tabulated, and returned to each participant for a second round, during which the participants are again asked to make an assessment regarding the same issue. The second time around the participants had the knowledge of what the other participants responded in the first round. The second round usually results in a narrowing of the range in assessments by the group, pointing to some reasonable middle ground regarding the issue of concern. The original Delphi technique avoided group discussion; but the Wideband Delphi technique (Boehm 1981) accommodated group discussion between assessment rounds. Two rounds of the Wideband Delphi survey for COSYSMO have been completed. The results are shown in the next section.

The second data collection effort involves steps 5 and 7. One completed project with an accurate count of systems engineering hours is considered one data point. These projects have been contributed by CSSE Affiliates that wish to have their application domains considered in the model. To date, eleven projects have been submitted to the COSYSMO repository for analysis.

Measurement reliability. An important experimental issue in field and quasi-experimental research designs is that of measurement reliability (Jarvenpaa, Dickson et al. 1985). This refers to the possible errors in measurement due to the accuracy of the measurement instrument. Surveys were used for both the Delphi and historical project data. Careful steps were taken to ensure that the design of the survey instrument followed the best practices in questionnaire design (Sudman and Bradburn 1982). Some of these are:

- Use of closed and open ended questions
- Knowledge questions to screen out respondents who lack sufficient information
- Consistent measurement scales for all questions of the same type
- Variability in polarity of the questions to avoid repetition
- Efficient use of space on the questionnaire

- Adequate level of difficulty of questions
- Easier questions at the beginning; difficult questions at the end
- Ample time to fill out questionnaire, typically 1 month
- Questionnaire is as short as possible while still covering the key points

These questionnaire features help the reliability of the data collection. One aspect outside of the researcher's control, however, is the administration of the questionnaire. Since these are sent to participants by e-mail there is no way to control its administration. Respondents, listed in Appendix C, complete the questionnaire at their own pace and in their own environment.

The most active organizations are also members of INCOSE and specialize in developing systems for military applications. One of the participants, Raytheon, has been extremely involved since the creation of the model and has implemented their own version of the model which they call *SECOST*.

Lessons Learned. Through the process of working with these organizations and refining the model definitions a number of useful lessons have been learned about collecting systems engineering data (Valerdi, Rieff et al. 2004). These were consolidated into eleven key findings as part of an exercise done in conjunction with some of the organizations listed in Appendix D. They include aspects such as scope of the model definitions, counting rules, data collection, and safeguarding procedures.

The research approaches, research designs, and lessons learned have played a significant role in the development of COSYSMO. In order to determine the predictive power of the model it has been validated through the use of statistical techniques. This effort represents step 5 of the seven step modeling methodology and is discussed in detail in the next section. One of the most human-intensive portions of this work was obtaining data from aerospace companies. The principal activities associated with the process of obtaining data from companies are illustrated in Figure 11.

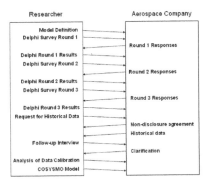

Figure 11 Data Handshaking

4.2. Threats to Validity & Limitations

Two types of threats to validity exist: controllable and uncontrollable. Great care has been taken to make sure COSYSMO is a useful model for systems engineering effort. Experimental design, however, is not perfect. Many external factors can affect the experiment and influence the overall result. This section attempts to identify the most significant threats to the validity of COSYSMO and outlines ways in which they were reduced. Most of these relate to *external* validity; the ability to generalize findings to other settings. Consequently, a COSYSMO user may ask: To what size or type of systems engineering projects can the model be generalized? The external validity of an experiment can be strengthened by describing the population to which the results will apply before the model is used. If a random systems engineering project is selected from a predetermined population (i.e, satellite ground stations) and COSYSMO yields an accurate estimate for systems engineering then the generalization can be made that systems engineering can be estimated with a certain degree of accuracy for satellite ground stations.

The data variability that is obtained from the data collection will determine how extensively the findings can be applied. If the model is calibrated from projects for military or defense applications, it cannot be claimed that the systems engineering CERs will hold true for other types of projects. The controllable threats to validity are:

1. Construct validity. The ensemble of size and cost drivers was carefully selected and tested to ensure that they were indeed adequate measures of systems engineering. They were also verified with heuristics, previous studies, and expert opinion.

67

2. Construct reliability. Counting rules and definitions were developed, with industry feedback, to guard against possible multiple interpretations and ensure consistent use of the drivers throughout different systems engineering domains.

3. Divergent definitions of Systems Engineering. An industry-accepted standard was adopted (EIA/ANSI 632) to aid in the identification of systems engineering activities through a baseline Work Breakdown Structure.

4. Experts. Expert sampling involves the assembling of a sample of persons with known or demonstrable experience and expertise in some area. Often, we convene such a sample under the auspices of a "panel of experts." There are actually two reasons one might do expert sampling. First, because it would be the best way to elicit the views of persons who have specific expertise. Second, because the expert sampling might be used to provide evidence for the validity of another sampling approach. For instance, the systems engineering drivers defined in COSYSMO need to be field tested by future users of the model. A panel of experts with experience and insight into systems engineering can examine the definitions and determine their appropriateness and validity. The advantage of doing this is that the drivers have significant practical relevance. The disadvantage is that even the experts can be, and often are, wrong. The responses in the Delphi survey came from "experts" in the field of systems engineering but the method used to administer the survey (e-mail) did not allow for screening of the survey respondents. To control for this, a set of questions was included at the beginning of the survey that asked respondents for their years of experience in systems engineering and/or cost estimation. The purpose of the question was to allow for the respondents to self select if they did not feel their experience was adequate to respond to the survey.

Other threats to validity exist which were identified but uncontrollable because they are often outside the range of control of the researcher. They may be the main source of error in the model since their impact is difficult to quantify:

1. Noisy data. Significant effort reporting differences and partial size information (i.e., size factor volatility and reuse) introduced some error in the data.

2. Nonresponse bias. Case study research is limited in its representativeness. The sample that is currently being used to calibrate COSYSMO is limited to the companies that have shown interest and made it possible to contribute systems engineering data or have been able to respond to the Delphi surveys. Because of the

narrow focus on only a few projects, there is a limitation to the model's generalizability because it only represents the CERs that are confirmed by the data set in use. Attempts have been made to include INCOSE and non-INCOSE companies that use systems engineering but only aerospace and defense companies affiliated with INCOSE were responsive. Moreover, these companies often exhibited a CMMI rating of 3 or higher; biasing the results towards high maturity organizations.

3. Sample Self-selection. This model involved the collection of data from six defense companies and the solicitation of inputs from experts employed at these companies which, in some cases, represent heterogeneous cultures due to the consolidation of the aerospace industry. Some of these companies have acquired portions of each other. As a result, a single aerospace company may reflect diverse cultural, productivity, and process standards inherited from all or portions of heritage companies it has acquired or merged with as shown in Table 46.

Table 46 Consolidation of Aerospace Companies

Company	Legacy
BAE Systems[4]	British Aircraft Corporation, Fairchild Systems, General Dynamics Electronics, Lockheed Martin Control Systems, Lockheed Martin Aerospace Electronics Systems, Marconi, Sanders
General Dynamics[5]	Allied-Signal, AT&T, Lucent, Digital System Resources, Lockheed Martin Power Control, Sylvania, Veridian, Western Electric/Bell Labs
Lockheed Martin[6]	Ford Aerospace, General Dynamics, General Electric Aerospace, Goodyear Aerospace, Gould Electronics, IBM Federal Systems, Lockheed, Loral, Martin Marietta, RCA, Vought, Unisys, United Space Alliance, Xerox Electro-Optical Systems
Northrop Grumman[7]	Aerojet, Grumman, Litton, Logicon, Newport News Shipbuilding, Northrop, TASC, Teledyne Ryan Aeronautical, TRW, Westinghouse
Raytheon[8]	E-Systems, Hughes, Texas Instruments

---

[4] http://www.na.baesystems.com
[5] http://www.generaldynamics.com
[6] http://www.lockheedmartin.com
[7] http://www.northropgrumman.com
[8] http://www.raytheon.com

4. Model will not work outside of calibrated range. The range of operation of COSYSMO is solely determined by the data that is used to calibrate it. Some users may attempt to use the model outside of it calibrated range which can lead to estimates with serious inaccuracies. As discussed in the previous chapter, no amount of disclaimer from the developer will keep the user from using the model to predict outside the region of the data.

5. Case study research is vulnerable to subjective biases. A project may be selected because of its dramatic, rather than typical, attributes; or because it is readily available. To the extent selective judgments exclude certain projects from the data set or assign a high or low value to a driver significance, or place them in one context rather than another, subjective interpretation is influencing the outcome. Moreover, only successful projects are reported for inclusion in COSYSMO. This makes the model biased towards successful projects because the unsuccessful projects do not collect or share data and are not included in the calibration.

6. Difficult to identify external variables. In quasi-experimental research it is difficult to identify all non-experimental project variables (i.e., competitive pressures, market trends, strategic advantages, etc.) and determine how to control or account for them.

7. Telescoping. The historical project data collection survey requires the respondent to go back in time and investigate the qualitative and quantitative parameters of the project. The quantitative parameters may be found in project documentation, if available. But the qualitative parameters will require a systems engineer or program manager to recollect a time frame in the past, referred to as telescoping. This technique has obvious drawbacks because individuals may not remember everything that happened in the past or they may be formulating their responses from secondary references. To overcome this, the focus is on obtaining historical project data on recently completed projects to increase the probability of accurate information and knowledgeable personnel with first hand experience.

8. Group Conformity. Solomon Asch's most famous experiments set a contest between physical and social reality. His subjects judged unambiguous stimuli – lines of different lengths – *after* hearing other opinions offering incorrect estimates. Subjects were very upset by the discrepancy between their perceptions and those of others and most caved under the pressure to conform: only 29% of his subjects refused to join the bogus majority. In a similar way, systems engineers could fall into the same trap

70

and agree with incorrect estimates based on what the group thinks is the correct choice.

With these threats in mind, a number of limitations also exist. The application domain profile of the application domain of Delphi survey respondents is shown in Figure 12. Nine out of forty (22%) participants who participated in Round 1 also took part in Round 2. To control for unfamiliarity of parametric models, follow up interviews were held with experts whose responses were outliers to get clarification on their answers. The average years of experience in software or systems engineering of survey participants was 18 years and the average years of experience in cost modeling was 6 years. Employees from fourteen different organizations participated in the survey but the majority (55%) of the participants were employees of Raytheon, Lockheed Martin, and Northrop Grumman. Proceed with caution if either (1) you are not one of the six companies that provided data, (2) your systems are outside of the size range for which the model is calibrated, and (3) your definition of systems engineering is not compatible with ANSI/EIA 632.

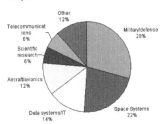

Figure 12 Application Domains of Delphi Participants

As shown in Figure 12 half of the participants selected "Military/defense" or "Space Systems" as their application domain which sheds light on the focus of the model. While the initial iteration of COSYSMO will most likely focus on these applications of systems engineering it will also provide useful results that other industries can compare themselves to.

Another important limitation of COSYSMO is its overlap with the well-known COCOMO II model. The danger with model overlap is that it can lead to unnecessary double-counting of effort because it is expected that systems engineering and software engineering are highly coupled in most organizations. The COCOMO II estimate of the software effort will surely account for the additional effort required by the additional testing;

71

at the same time, the COSYSMO effort will account for additional test development and management since the systems engineers are required to perform additional validation and verification of the system. Either model can account for this effort based on how users wish to allocate the testing activity. Each organization's unique relationship between these two disciplines needs to be reconciled when using COSYSMO and COCOMO II together. One approach for accomplishing this is to examine the Work Breakdown Structures of each discipline.

COSYSMO uses the WBS defined in EIA/ANSI 632 while COCOMO II uses the one defined in MBASE/RUP. The two models are more likely to demonstrate overlap in effort in the case of software-intensive systems. Table 47 shows the activities that could potentially overlap when using both models during an effort estimation exercise. The numbers in the cells represent the typical percentage of effort spent on each activity during a certain phase of the software development life cycle as defined by COCOMO II. Each column adds up to 100 percent.

Table 47 COCOMO II and COSYSMO Overlaps

Project Stage	Software Development			
	Inception	Elaboration	Construction	Transition
Management	14	12	10	14
Environment/CM	10	8	5	5
Requirements	38	18	8	4
Design	19	36	16	4
Implementation	8	13	34	19
Assessment	8	10	24	24
Deployment	3	3	3	30

COCOMO II
COSYSMO
COCOMO II/COSYSMO overlap

The checkered cells indicate the COCOMO II/COSYSMO overlap activities that may be double counted when using the models simultaneously. The gray cells indicate the systems engineering activities that are estimated in COSYSMO. The exact amount of effort being double counted will vary for each organization based on the way they define systems engineering relative to software engineering.

72

5. Results and Next Steps

5.1. Delphi Results

A Delphi exercise was conducted to reach group consensus and validate initial findings. The Wideband Delphi technique has been identified as being a powerful tool for achieving group consensus on decisions involving unquantifiable criteria (Boehm 1981). It was used it to circulate the initial findings and reach consensus on the parametric ratings provided by experts. The cumulative experience of the experts in the second round includes over 200 years of cost estimation and 600 years of systems engineering. Part of the Wideband Delphi technique involves face-to-face meetings to review the results of the previous round and discuss any possible changes. Eleven COSYSMO Wideband Delphi meetings took place between March 2002 and March 2004.

Part of the Delphi process involved multiple distributions of the surveys to arrive at the values that experts could converge on. The purpose of the survey was to (1) reach consensus from a sample of systems engineering experts, (2) determine the distribution of effort across effort categories, (3) validate the drivers of systems engineering size, (4) identify the cost drivers which have the most influence on effort, and (5) help the refinement of the scope of the model elements.

Each size driver reflects the range of impact and variation assigned by the experts during the refinement exercises. The group of size drivers includes a volatility factor that accounts for the amount of change that is involved in the four factors. For example, it can be used to adjust the number of requirements should they be ill-defined, changing, or unknown at the time the estimate is being formulated.

Delphi Results. After three Wideband Delphi meetings, COSYSMO was refined to reflect what INCOSE and other systems engineering organizations felt the most significant size and cost drivers were. The relative weights for "Easy", "Nominal", and "Difficult" levels are presented in Table 48.

Table 48 Relative Weights for Size Drivers from Delphi Round 3

	Easy	Nominal	Difficult
# of System Requirements	0.5	1.0	5
# of Major Interfaces	1.7	4.3	9.8
# of Critical Algorithms	3.4	6.5	18.2
# of Operational Scenarios	9.8	22.8	47.4

The *# of Systems Requirements* driver was kept as the frame of reference to the other three size drivers. A graphical representation of these results is provided in Figure 13.

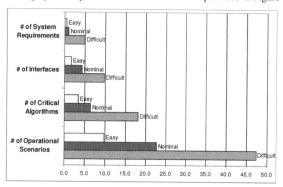

Figure 13 Relative Weights for Size Drivers from Delphi Round 3

One interpretation of the size drivers is that a "Difficult" systems requirement requires five times the effort relative to a "Nominal" one. Additionally, a "Difficult" operational scenario, the most influential size driver, requires forty seven times the effort of one "Nominal" system requirement, assuming a linear scale factor. The fourteen cost drivers and their respective rating scales are shown in Table 49. The individual values for all the applicable rating levels are provided. The nominal ratings for all the drivers are always 1.0, but the polarity of the ratings depends on the definition of the driver.

Table 49 Rating Scale Values for Cost Drivers from Delphi Round 3

	Very Low	Low	Nominal	High	Very High	Extra High	EMR
Requirements Understanding	1.87	1.37	1.00	0.77	0.60		*3.12*
Architecture Understanding	1.64	1.28	1.00	0.81	0.65		*2.52*
Level of Service Requirements	0.62	0.79	1.00	1.36	1.85		*2.98*
Migration Complexity			1.00	1.25	1.55	1.93	*1.93*
Technology Risk	0.67	0.82	1.00	1.32	1.75		*2.61*
Documentation	0.78	0.88	1.00	1.13	1.28		*1.64*
# and diversity of installations/platforms			1.00	1.23	1.52	1.87	*1.87*
# of recursive levels in the design	0.76	0.87	1.00	1.21	1.47		*1.93*
Stakeholder team cohesion	1.50	1.22	1.00	0.81	0.65		*2.31*
Personnel/team capability	1.50	1.22	1.00	0.81	0.65		*2.31*
Personnel experience/continuity	1.48	1.22	1.00	0.82	0.67		*2.21*
Process capability	1.47	1.21	1.00	0.88	0.77	0.68	*2.16*
Multisite coordination	1.39	1.18	1.00	0.90	0.80	0.72	*1.93*
Tool support	1.39	1.18	1.00	0.85	0.72		*1.93*

For example, the *Requirements Understanding* driver is worded positively since there is an effort savings associated with high or very high understanding of the requirements. This is indicated by multipliers of 0.77 and 0.60, respectively representing a 23% and 40% savings in effort compared to the nominal case. Alternatively, the *Technology Risk* driver has a cost penalty of 32% for "High" and 75% for "Very High". Not all rating levels apply to all of the drivers. Again, it is a matter of how the drivers are defined. The *Migration Complexity* driver, for example, only contains ratings at "Nominal" or higher. The rationale behind this is that the more complex the legacy system migration becomes, the more systems engineering work will be required. Not having a legacy system as a concern, however, does not translate to a savings in effort. The absence of a legacy system is the nominal case which corresponds to a multiplier of 1.0.

The cost drivers are compared to each other in terms of their range of variability, or Effort Multiplier Ratio. The EMR column is representative of an individual driver's possible influence on systems engineering effort. The cost drivers are presented in order of EMR value in Figure 14. The four most influential cost drivers are: *Requirements Understanding, Level of Service Requirements, Technology Risk,* and *Architecture Understanding*. The least influential, *Documentation, # of Installations, Tool Support,* and *# of Recursive Levels in the Design* were kept because users wanted to have the capability to estimate their impacts on systems engineering effort. The relatively small influence of these four drivers does not mean that the model users felt they were insignificant. Their presence gives users the ability to quantify their impact on systems engineering. This is what some researchers refer to as the difference between statistical significance and practical significance (Isaac and Michael 1997).

In the quest for obtain statistically significant findings, relevant factors may be overlooked. Are quality issues and testing issues significant to systems engineering? Is their influence important enough to be practical and included in COSYSMO? Is the data for these drivers important enough to be worth the effort to obtain it? Even when these practical matters are settled, there are valuable considerations of social and psychological nature that can override cost driver choices based solely on statistical significance. In fact, debates continue to take place regarding the addition of other cost drivers into the model. These debates naturally will go on as more contributors join the COSYSMO development effort. For now, the belief is that the most significant drivers have been identified and COSYSMO will be easily validated by the drivers' relationships to systems engineering effort. As shown in Appendix F, no two cost drivers are correlated higher than 0.6; validating that a reasonably orthogonal set has been identified.

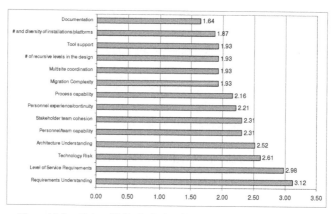

Figure 14 Cost Driver EMRs in Order of Influence from Delphi Round 3

5.2. Model Verification

Much energy has been spent on defining the inputs, outputs, counting rules, definitions, and level of detail required for this model. The adage "A problem well defined is a problem half solved" reiterates the importance of clear definitions and model scope. In a general sense, this model has been developed in the same way as deliverable and complex systems are, following an iterative systems engineering approach.

The preceding chapters provided a description of the problem and an approach. The next logical step is to shift the focus to model verification. The process of model verification is twofold: first, a series of statistical tests are performed. Second, feedback is obtained from end users to determine the impact of the model. Specifically, step 5 of the model development methodology was performed as described in (Baik, Boehm et al. 2002) using the procedure outlined in (Chulani, Boehm et al. 1999) on the COSYSMO datapoints. This section provides only the steps that were performed to analyze the data, not the data itself since the source data provided by organizations is protected for reasons of industrial security. Once data is obtained and a regression model is defined, a number of useful diagnostics tests are helpful in validating the regression model (Snee 1977). These include:

- Comparison of model predictors and coefficients with physical theory
- Data splitting to obtain an independent measure of the model prediction accuracy
- Outlier/residual analysis

The form of the model follows the Ordinary Least Squared (OLS) criterion because it seeks to find a simple linear regression mean function to represent the relationship between the 18 systems engineering drivers, the independent variables, and systems engineering effort, the dependent variable. The OLS approach has four inherent assumptions (Griffiths, Hill et al. 1993): (1) there is a lot of data available, (2) no outliers exist, (3) predictor variables are not correlated, and (4) predictors are either all continuous or all discrete. In an ideal case all of these assumptions would be true, but in reality it is difficult to find a dataset with one or more of these characteristics. Steps have been taken to meet these assumptions as much as possible without interfering with the systems being studied. The first assumption was addressed by collecting data from 42 projects and 70 experts. The second assumption was not realistic because some projects were indeed outliers and had to be removed from the dataset due to the fact that they addressed systems engineering size differently than other programs. The third assumption was reasonably addressed which can be seen in Appendix F. The fourth assumption was met by making all predictors continuous.

77

5.2.1. Statistical Tests

COSYSMO can be characterized by a multiple regression model where the response is Person Months (PM) and the predictors are the 18 drivers that have an influence on systems engineering effort. This linear function is estimated from the data using the ordinary least squares (OLS) approach as defined by (Cook and Weisberg 1999). The multiple regression model can be written in the form:

$$\text{Equation 7} \qquad y_t = \beta_0 + \beta_1 x_{t1} + \ldots + \beta_k x_{tx} + \varepsilon_t$$

Where $x_{t1} \ldots x_{tk}$ represent the values of the predictor variables for the t_{th} observation, $\beta_0 \ldots \beta_k$ are the coefficients estimated via the OLS regression, ε_t is the error term, and y_t is the response variable for the t_{th} observation. Based on the normalizing transformations needed to express linear relationships in the model, logarithmic transformations are applied to the dependent and independent variables of the equation yielding:

$$\text{Equation 8}$$
$$\ln(SE\_HRS) = \beta_0 + \beta_1 \cdot \ln(S_1) + \ldots + \beta_4 \cdot \ln(S_4) + \beta_5 \cdot \ln(EM_1) + \ldots + \beta_{18} \cdot \ln(EM_{14})$$

The assumption of the logarithmic transformation is based on experience of inspecting software engineering data from COCOMO II. Systems engineering data should behave in a similar function and require the same transformation to be normalized. Systems Engineering hours, denoted as *SE_HRS*, was used in order to avoid the discrepancy between different Person-Month standards. The COCOMO suite of models use 152 hours per Person Month which can be adjusted depending on user preference. The four size and fourteen cost predictors are listed in Table 50.

Table 50 COSYSMO Predictor Descriptions

Predictor	Term	Description
S_1	log(REQ)	# of System Requirements
S_2	log(INTF)	# of Major Interfaces
S_3	log(ALG)	# of Critical Algorithms
S_4	log(OPSC)	# of Operational Scenarios
EM_1	log(RQMT)	Requirements Understanding
EM_2	log(ARCH)	Architecture Understanding
EM_3	log(LSVC)	Level of Service Requirements
EM_4	log(MIGR)	Migration Complexity
EM_5	log(TRSK)	Technology Risk
EM_6	log(DOCU)	Documentation to match lifecycle needs
EM_7	log(INST)	# and diversity of installations/platforms
EM_8	log(RECU)	# of recursive levels in the design
EM_9	log(TEAM)	Stakeholder Team Cohesion
EM_{10}	log(PCAP)	Personnel/Team Capability
EM_{11}	log(PEXP)	Personnel Experience/Continuity
EM_{12}	log(PROC)	Process Capability
EM_{13}	log(SITE)	Multisite Coordination
EM_{14}	log(TOOL)	Tool Support

Once the data was collected and entered into a repository, the following multiple regression diagnostic tests were performed using the *Arc* software, a freely-available academic statistics package (Cook and Weisberg 1999):

- Model significance/F-test. A series of F-tests were performed to compare the full model to it variations. This procedure compares the difference between the Residual Sum of Squares (RSS) of the null hypothesis, or full model, and the RSS of the alternative hypothesis, or reduced model. Also considered are the differences between the degrees of freedom of the two models and the Mean Square Error (MSE) of the alternative hypothesis. This can be represented as:

$$\text{Equation 9} \qquad F = \frac{(RSS_{NH} - RSS_{AH})/(df_{NH} - df_{AH})}{MSE_{AH}}$$

Large values of F provide evidence against the null hypothesis and in favor of the alternative hypothesis. Significance levels for this test can be obtained in statistics textbooks (Cook and Weisberg 1999). The F-values for the different models tested are shown in Table 52.

- Correlation Matrix. The correlation matrix can serve as an indicator of both strong and weak correlations between predictors. Since the desire is to have a model with truly independent and orthogonal predictors, the correlations above 0.66 are flagged as being strong and possible candidates for elimination. This criterion was previously used in COCOMO II (Chulani, Boehm et al. 1999). See Appendix F for numerical results.

- Sensitivity analysis. The introduction of new data points to the model were tested for influence on the significance of the model predictors. The t-values and p-values provided information about the influence of individual predictors on the mean function. The t-values are the ratio between the estimate and its corresponding standard error, where the standard error is the square root of the variance. It can also be interpreted as the signal-to-noise ratio associated with the corresponding predictor variable. Hence, the higher the t-value, the stronger the signal or statistical significance of the predictor variable. Typically, a high t-value is approximately 3.0 or 4.0, indicating statistical significance for predictor variables. The p-values are an indication of evidence that the probability of observing a value of the statistic is high or low. Values less then 0.1 are an indication of strong predictive influence on the mean function. The t-values and p-values for the final model are shown in Appendix H.

- Stepwise Regression. Two algorithms widely available in most statistics packages, and in this case implemented in *Arc*, are backward elimination and forward elimination. These are useful in evaluating submodels by sequentially adding or removing predictor terms and comparing the results. Stepwise regression algorithms do not guarantee finding optimal submodels, although the results obtained in this approach are useful in determining candidates for elimination. This approach was previously used in COCOMO II (Baik 2000) and for this data set helped arrive at the reduced form of the model. Forward selection was a better indicator of the best arrangement of predictors for COSYSMO because of the particular behavior of the data set and small number of degrees of freedom.

5.2.2. Model Parsimony

As stated earlier, one of the key objectives for COSYSMO is to avoid the use of highly redundant parameters as well as factors which make no appreciable contribution to the results. In order to achieve this, four variations of the full model were tested to arrive at the

80

final model that met all of the accuracy, parsimony, constructiveness, and simplicity objectives previously defined.

Full model. The complete set of parameters described in sections 3.2 and 3.3 serve as the baseline model. They are listed again in Table 50 in logarithmic scale. Equation 8 can be rewritten as:

Equation 10

$$\log[SE\_HRS] = \log[REQ] + \log[INTF] + \log[ALG] + \log[OPSC] + \\ \log[RQMT] + \log[ARCH] + ... + \log[TOOL]$$

The advantage of starting with a baseline model that contains all of the model parameters is that tests for pair wise orthogonality can be performed. The results of this analysis, found in Appendix F, show that no two cost drivers are correlated more than 0.66, indicating that they represent a parsimonious set. Some of the four size drivers are correlated as high as 0.64 which was expected since system requirements are often related to interfaces, algorithms, and operational scenarios.

Reduced model. To address the issue of correlated size drivers and reduce the number of predictors in the model, the four size drivers were combined into a single predictor called *Size*. Equation 10 can be written as:

Equation 11

$$\log[SE\_HRS] = \log[SIZE] + \log[RQMT] + \log[ARCH] + ... + \log[TOOL]$$

It was found that the reduced model containing the *Size* parameter had a higher accuracy than the full model. In other words, the combination of the size drivers had a higher explanatory power together rather than individually. Combining the four size driver into one parameter also increased the degrees of freedom.

Adjusted model. The systems engineering hours for the projects reported by participants were not provided uniformly. Only 13 of the 42 projects provided effort for systems engineering activities during the entire development life cycle while the rest provided effort for only some of the life cycle phases defined in the model. As a result, 29 of the 42 projects had to be normalized to fit the four phases of the development life cycle that were most commonly reported: Conceptualize, Develop, OT&E, and Transition to Operation. The typical distribution of effort across these phases was obtained by surveying industry experts familiar with COSYSMO. These results, shown in Table 51, were derived from the more detailed results shown in Appendix B. It should be noted that the standard deviations associated with the results are relatively large indicating disagreement between experts on the typical distribution of systems engineering effort across the four phases. Nevertheless, this profile of systems engineering effort across the four life cycle phases was useful in adjusting the reported effort on projects.

Table 51 Systems Engineering Effort Distribution % Across ISO/IEC 15288 Phases

Conceptualize	Develop	Operational Test & Eval	Transition to Operation
23	35	28	14

Projects that only reported hours for the development phase, in theory, were leaving out 65% of the effort had the project covered the first four life cycle phases. As a result, the hours reported were adjusted to add the missing 65% in order to normalize the data point.

Conceptually, COSYSMO can be used to estimate systems engineering effort for the entire life cycle. This can be done by using a similar approach used in COCOMO II which estimates effort as a function of annual change traffic. Since the Operate, Maintain, or Enhance; and Replace or Dismantle phases are often assigned resources by "level of effort" it makes sense to estimate systems engineering resources based on how much the system changes once it is fielded. The four size drivers can be tracked in terms of annual change traffic and the cost drivers can be reassessed based on the changing conditions of the program.

The normalized effort is denoted as *SE_HRS_ADJ* to reflect the adjustment applied to the reported hours. Equation 11 is therefore rewritten as:

Equation 12

$$\log[SE\_HRS\_ADJ] = \log[SIZE] + \log[RQMT] + \log[ARCH] + \ldots + \log[TOOL]$$

The model shown in Equation 12 yielded a better representation of the systems engineering hours spent on projects. The relationship between *SIZE* and *SE_HRS_ADJ* for the 42 data points is shown in Figure 15. The 8 data points that were removed from the analysis are shown in black while the rest of the projects are in hollow circles.

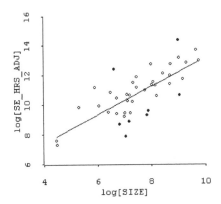

Figure 15 Size Versus Adjusted Systems Engineering Hours

Part of the data analysis included the elimination of points believed to be outliers. The first test for outliers was a measure of productivity, shown in Equation 13.

Equation 13

PRODUCTIVITY = SIZE/SE_HRS_ADJ

It was determined that projects above a productivity of 0.14 provided their requirements count at a different level of decomposition than the rest of the projects. As a result, the productivity for these particular six projects was much higher than the rest of the data set; identifying a reduced domain for which a model would be more accurate. The productivities for the 42 projects, with the 6 projects highlighted on the far right, are shown in Figure 16.

83

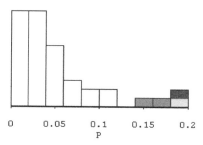

0   0.05   0.1   0.15   0.2
P

Figure 16 Productivity Histogram for 42 projects

In addition to the 6 projects eliminated due to extremely high productivity, 2 projects were also removed from the calibration domain due to a high degree of adjusted effort (more than 30%) across the systems engineering life cycle.

Final Model. With a smaller number of data points, a smaller number of aggregated parameters was considered. The cost drivers were clustered into groups in section 3.3 based on common themes. These five groups were determined to be logical groupings of cost drivers that addressed similar issues dealing with systems engineering. This resulted in the model shown in Equation 14.

Equation 14

$$\log[SE\_HRS\_ADJ] = \log[SIZE] + \log[UNDERSTANDING] + \\ \log[COMPLEXITY] + \log[OPERATIONS] + \\ \log[PEOPLE] + \log[ENVIRONMENT]$$

Where
UNDERSTANDING = REQU * ARCH * TEAM * PEXP
COMPLEXITY = LSVC * TRSK * RECU * DOCU
OPERATIONS = INST * MIGR
PEOPLE = PCAP * PROC
ENVIRONMENT = SITE * TOOL

The model in Equation 14 provided additional degrees of freedom and a higher F-value. However, not all six predictors met the p-value criterion of 0.1. See regression results for this model in Appendix H. Nevertheless, this model was used as the final calibrated version because it contained all of the parameters in the model in aggregated form.

84

Reduced model. A reduced form of the model was developed which includes predictors whose p-values were less than 0.1. These are shown in Equation 15. The use of forward selection was helpful in identifying the best combination of parameters from the aggregated model. However, this model has limited use since it includes less than half of the model parameters.

Equation 15

$$\log[SE\_HRS\_ADJ] = \log[SIZE] + \log[COMPLEXITY] + \log[PEOPLE]$$

Table 52 provides a comparison of model performance for the five permutations of COSYSMO.

Table 52 Comparison of Model Performance

Model iteration	predictors	R-squared	Degrees of freedom	F-value
Full model	18	0.64	20	1.98
Reduced Model	15	0.63	23	2.66
Adjusted model	15	0.77	23	5.36
Final model **(used for Bayesian calibration)**	6	0.81	27	20.44
Reduced model	3	0.74	29	29.14

The most representative model was decided to be the Final model since it had a reasonable number of degrees of freedom while offering the ability to derive the original size and cost drivers. Different PRED accuracy levels (Conte et al 1986) tested on this version of the model and are shown in Table 53.

Table 53 Model Accuracy of Delphi Based Model

	Accuracy
PRED(.20)	19%
PRED(.25)	22%
PRED(.30)	30%

5.2.3. Bayesian Approximation

A Bayesian approximation was performed on the Final model since it contains, at least in aggregated form, all of the original parameters in the model. The results yield the final calibrated model which, by updating Equation 2, can be written as:

$$PM_{NS} = A \cdot (Size)^E \cdot \prod_{i=1}^{n} EM_i$$

A = 38.55

Size = weights in Table 18

E = 1.06

n = number of cost drivers (14)

EM = multipliers in Table 19

Equation 16 Final Bayesian Calibrated Model

A benefit of using the Bayesian method is that it allowed for expert opinion from the Delphi survey to influence the calibration derived from the historical data. Negative coefficients are representative of contradictory results and are common when limited amounts of data exist (Chulani et al 1999). The updated weights and ratings scale multipliers from the Bayesian calibration are provided in Table 54 and Table 55.

Table 54 Relative Weights for Size Drivers for Bayesian Calibrated Model

	Easy	Nominal	Difficult
# of System Requirements	0.5	1.0	5
# of Major Interfaces	1.1	2.8	6.3
# of Critical Algorithms	2.2	4.1	11.5
# of Operational Scenarios	6.2	14.4	30

Table 55 Bayesian Calibrated Rating Scale Multipliers

	Very Low	Low	Nominal	High	Very High	Extra High	EMR
Requirements Understanding	1.85	1.36	1.00	0.77	0.60		3.08
Architecture Understanding	1.62	1.27	1.00	0.81	0.65		2.49
Level of Service Requirements	0.62	0.79	1.00	1.32	1.74		2.81
Migration Complexity			1.00	1.24	1.54	1.92	1.92
Technology Risk	0.70	0.84	1.00	1.32	1.74		2.49
Documentation	0.82	0.91	1.00	1.13	1.28		1.56
# and diversity of installations/platforms			1.00	1.23	1.51	1.86	1.86
# of recursive levels in the design	0.80	0.89	1.00	1.21	1.46		1.83
Stakeholder team cohesion	1.50	1.22	1.00	0.81	0.66		2.27
Personnel/team capability	1.48	1.22	1.00	0.81	0.66		2.28
Personnel experience/continuity	1.46	1.21	1.00	0.82	0.67		2.18
Process capability	1.46	1.21	1.00	0.88	0.77	0.68	2.15
Multisite coordination	1.33	1.15	1.00	0.90	0.80	0.72	1.85
Tool support	1.34	1.16	1.00	0.85	0.73		1.84

An example estimate using the final calibrated COSYSMO model is provided in Appendix E. The accuracy of the model on the set of 34 projects (6 eliminated because of high productivity and 2 eliminated due to small percentage of effort reported) is provided in Table 56.

Table 56 Model Accuracy of Bayesian Calibrated Model

	Accuracy
PRED(.20)	19%
PRED(.25)	27%
PRED(.30)	41%

A slight accuracy improvement can be seen from the Delphi-based model in Table 53 to the Bayesian calibrated model in Table 56. In addition, the use of the Bayesian calibrated model enabled the elimination of negative coefficient in PEOPLE parameter shown in the regression results in Appendix H.

5.2.4. Stratification by Organization

Using the final model, a set of local calibrations were done for each individual organization to determine the effect on model accuracy. The results were obtainable only for the companies that provided more than four data points.

Table 57 Model Accuracy by Organization

Organization	N	R-squared	PRED(30)
1	10	0.94	70%
2	7	0.59	43%
3	10	0.62	50%
All	**27**	-	**56%**

The results in Table 57 validate the model accuracy posed in Hypothesis #4; PRED(30) at 50%. In some cases, the prediction accuracy of the model is improved even further when the organization's disciplined data collection has uniform counting rules that can ensure consistency across programs. Regardless of the results provided by the statistical validation, limitations of the model should be considered when the prediction accuracy of the model is determined. Limitations were provided in previous sections as were their implications on model validity.

In addition to the rigorous quantitative analysis of the model, a qualitative analysis was performed to determine the impact of the model on each organization that provided data. COSYSMO supporters were enthusiastic about the model's influence in the way their organizations thought about systems engineering cost. Out of 8 participants queried, 7 of them responded that COSYSMO greatly improved their ability to reason about systems engineering cost. Example testimonials include:

- Model helps answer the call for improvement on systems engineering revitalization and ensuring adequate SE resources identified in the Young Panel report for DoD Space Systems
- The size and effort drivers will allow the user of the model to best describe the project and to perform sensitivity analysis to try to optimize the SE application and team aspects of the project

- COSYSMO answers the 'CMMI mail' regarding a key requirement for Maturity Level 2: Costing by Attributes ---- specifically, counts of System Requirements, System Interfaces, Critical Algorithms, and Operational Scenarios as scaled by the environmental (team and application) cost drivers.

5.3. Conclusion

A new type of systems engineering cost model was presented and validated through industry data. An objectively reduced subset of 42 industry projects from 6 companies was used to calibrate the model in reduced form which produced a PRED(30) accuracy of 50%. Together with counting rules and driver definitions, the model provides a way for systems engineers to reason about their cost decisions and presents a new vehicle for managers to approach systems engineering strategy via size and cost metrics.

Restatement of hypotheses. The five hypotheses stated in Section 1.3 were tested to determine their validity.

H#1: A combination of the four elements of functional size in COSYSMO contributes significantly to the accurate estimation of systems engineering effort.

The criterion used was a significance level less than or equal to 0.10 which translates to a 90% confidence level that these elements are significant. This hypothesis was <u>supported</u> with the reduced form of the model.

H#2: An ensemble of COSYSMO effort multipliers contribute significantly to the accurate estimation of systems engineering.

The same significance level of 0.10 was used to test this hypothesis which was <u>supported</u> with the reduced form of the model.

H#3: The value of the COSYSMO exponent, E, which can represent economies/diseconomies of scale is greater than 1.0.

For some organizations, the exponent was as low as 0.83 and as high as 1.31, as a result this hypothesis was <u>partially supported</u>. The average value for E was 1.06.

H#4: There exists a subset of systems engineering projects for which it is possible to create a parametric model that will estimate systems engineering effort at a PRED(30) accuracy of 50%.

89

This hypothesis was supported with the subset of projects defined as the projects from organizations with more than 4 solid data points with systems engineering productivity below 0.14 size units per adjusted SE hours.

H#5: COSYSMO makes organizations think differently about Systems Engineering cost.

This hypothesis was strongly supported. Seven out of eight responded with resounding support within the COSYSMO pioneer community as evidenced by the testimonials in section 5.2.4; insufficient data elsewhere.

5.3.1. Contributions to the Field of Systems Engineering

This work has a number of significant contributions to the field of systems engineering. These contributions have implications for researchers, practitioners, and educators.

1. Development of a parametric Systems Engineering cost model. Practitioners will benefit from the creation of COSYSMO because it is the first cost estimation model that provides systems engineering effort estimation. The identification of significant size and cost drivers for systems engineering can also serve as a risk management list for projects.

2. Systems Engineering Sizing. The development of counting rules for systems engineering using non-software metrics such as lines of code or function points is a significant contribution to the field of parametric cost modeling.

3. Operationalization of cost drivers and rating scales. The list of 14 cost drivers, their definitions, and their Cost Estimating Relationships provide useful teaching tools for systems engineering educators. Results show that these cost drivers are independent of each other and are good indicators of systems engineering complexity. A list of the cost drivers in order of influence was also generated to help people identify the factors that have the largest influence on cost.

4. SE as an independent discipline. Organizations such as INCOSE will benefit from COSYSMO because it helps justify the value of systems engineering in organizations. Researchers in the area of parametrics can also build from this work to develop additional CERs that are independent of software or hardware development.

5. Development of the first systems engineering focused model. COSYSMO can also help efforts currently underway to develop frameworks and theories related to systems engineering. One such example is the Grand Unified Theory of Systems

Engineering (GUTSE) developed by George Friedman (Friedman 2003) which implies the future existence of a reliable cost model but currently does not include such a model.

6. Institutional memory. The development of COSYSMO has captured over 500 person years of experience in systems engineering experience from over a dozen heterogeneous organizations. This baseline model can help researchers and educators develop case studies that can further the understanding of systems engineering.

7. Convergence. Industry experts were orchestrated to determine Delphi weights, driver definitions, counting rules, model form, and model scope. Convergence on these values demonstrates industry collaboration through the facilitation of academic researchers.

8. Evidence of diseconomy of scale. Empirical evidence verifies the hypothesis that systems engineering effort exhibits a diseconomy of scale.

9. Systems engineering effort profile. Through expert opinion, an effort profile was developed using the fundamental processes in ANSI/EIA 632 to define the model scope and distribute effort estimates across activities.

In addition to the practitioners that have benefited from COSYSMO, the following educational institutions have incorporated it into their systems engineering syllabus: University of California San Diego, George Mason University, Defense Acquisition University, and Southern Methodist University.

5.3.2. Areas for Future Work

The estimation of systems engineering cost is a complex but much needed process. The acquisition community continues to demand more accurate quantification of the SE work that is performed on hardware and software systems. In order to realize more sophisticated methods for estimating SE effort, the models should be based on empirical data. Rather than using rules of thumb, models should be data-driven. Instead of having limited relevance only to standard life cycles, they should be more flexible. Further work is required to make the task of managing SE more quantitative.

This section is divided into two parts to reflect the areas for future work that are needed in the short term and the long term. The short term areas are ones that will need to be addressed as extensions of this work while the long term are ones that can provide future opportunities for research in this area.

The process of developing COSYSMO has produced a number of useful Cost Estimating Relationships (CERs) that aid in estimating systems engineering effort. These CERs are accompanied by a set of rules that help guide the user in defining the scope of COSYSMO and how it relates to their organization. One of the challenges of creating a generalizable model is that it has to be flexible enough to work in different application domains. These domains operate differently by the nature of the customers they serve, the systems they build, and the environment in which their systems must operate. COSYSMO constituents represent a diverse set of users that represent an additional level of complexity because of their heterogeneous nature. This diversity has introduced the following challenges that need to be resolved:

1. Effects of volatility and reuse. These effects were deferred because of the difficulty in gathering data to calibrate them. Future work should incorporate these relevant parameters into the model.

2. Requirements counting. The different levels of requirements decomposition that are used by organizations was previously presented. More specific counting rules need to be developed to ensure consistency in this area.

3. The SEMP. Effects of the Systems Engineering Management Plan (SEMP) is currently not included in COSYSMO, however, it plays a significant role in the effectiveness of systems engineering organizations. For the time being, it has been included as part of the *Process Capability* cost driver. Future exploration is required to define the best rating scale that captures its influence on systems engineering effort.

4. Drivers with multiple viewpoints. The consolidation of the model discussed in Chapter 8 brought about an overloading of some cost drivers. This was caused by the merging of two or more drivers into one resulting in the introduction of multiple viewpoints for some drivers. Drivers such as *Technology Risk, Multisite Coordination, Migration Complexity,* and *# and Diversity of Installations & Platforms* have two or more aspects of systems engineering embedded in their definitions. These need to be simplified or rolled into separate cost drivers.

5. CMMI & Key Process Areas. The influence of CMMI as a cornerstone of this model was highlighted in section 3. However, not every organization has adopted the CMMI framework in its processes. As a result, an alternative method for measuring the process capability of an organization needs to be developed through the use of Key Process Areas (KPAs). Moreover, some confusion exists in situations where

multiple organizations with different levels of maturity are operating under the same contract. This situation is difficult to quantify from the COSYSMO standpoint and needs to be resolved.

6. Technology Readiness Levels. Another difficulty in quantifying a cost driver occurs when multiple technologies with different maturity levels and different time horizons have an effect on the systems engineering organization(s).

7. EIA/ANSI 632 activities. The menu of EIA/ANSI 632 activities listed in Chapter 3 represents an industry standard which is not entirely applicable in practice. Further investigation is needed to determine which systems engineering activities in EIA/ANSI 632 are more likely to be performed by organizations using COSYSMO.

8. Compatibility with other models. The MBASE framework presented includes several other product models that are closely related to systems engineering. Software engineering activities estimated by COCOMO and system of system software integration activities estimated by COSOSIMO need to be clearly delineated to ensure minimal overlap or ways to account for their overlap with COSYSMO.

9. Need more data. The list of organizations that have contributed data provided enough data to calibrate an initial model, but more data is needed in order to perform more tests of significance on all the model parameters.

10. Need more commercial company involvement. The same list of aforementioned organizations indicates a heavy bias toward aerospace and military projects. More commercial companies' data needs to be obtained and incorporated into the model calibration.

Long Term Areas. In order for COSYSMO to be a useful tool for the systems engineering community it must be adaptable to a situation of interest. The initial release of the model will have an industry baseline calibration that is representative of the CERs and driver definitions at the industry level. The model will be much more useful to individual organizations if it is calibrated for their use. This and other adaptations to the model are long term opportunities for future research.

1. COSYSMO Tailoring. Customizing the model to meet the needs of organizations is necessary to maximize the model's predictive ability. Local calibrations should be done and, if possible, product line calibrations for a specific family of products and customer calibrations for particular operational environments.

2. Additional driver definitions. Part of the customization process may involve the development of additional cost or size drivers. This alternative should especially be investigated by commercial companies that participated in the initial calibration.

3. Maintenance model. The CERs for systems engineering may behave differently in the maintenance phase of the project life cycle. This could serve as an opportunity to develop a COSYSMO maintenance model.

4. Tool comparison matrix. The development of a tool comparison matrix would be of value to enable the selection of the most appropriate cost estimating tool for a particular project. Such a comparison would establish uniform definitions and factors that would encompass the breadth of how COCOMO, COSYSMO, and COSOSIMO assess systems engineering. Such a comparison would facilitate identification of areas lacking sufficient detail and establish understanding to appropriately quantify systems engineering. These findings would suggest common areas of enhancement and development for additional research.

5. Risk Analyzer. Since COSYSMO provides a single point estimate, it would benefit from a risk analysis tool that determines the amount of risk associated with a given estimate.

6. Additional behavioral implications. The process of effort estimation involves unique skills and knowledge about how people work together and how organizations interact with each other. It would be interesting to investigate how other types of knowledge can affect the process of cost estimation. Some possible areas of investigation are: codified knowledge, Herbert Simon's bounded rationality (what is in and what is out), Eric von Hippel's sticky information (people's ability to remember things), Eugene Ferguson's idea of the mind's eye (aesthetic knowledge), and group think.

As industry and academia collaborate to develop and validate a tool to more accurately forecast SE resources, more opportunities for research are created. These areas could potentially change the way cost estimation models are developed and used.

References

Abts, C., Boehm, B. W. and Bailey-Clark, E. (2001). COCOTS : A Software COTS-Based System (CBS) Cost Model - Evolving Towards Maintenance Phase Modeling. Proceedings of the Twelfth Software Control and Metrics Conference, London, England.

Ahern, D. M., Clouse, A. and Turner, R. (2004). CMMI Distilled - A Practical Guide to Integrated Process Improvement, Addison-Wesley.

Albrecht, A. J., Gaffney, J., (1983), "Software function, source lines of code, and development effort prediction: A software Science validation," IEEE Transactions on Software Engineering, **Vol. SE-9**, pp. 639-648.

ANSI/EIA (1999). ANSI/EIA-632-1988 Processes for Engineering a System.

ANSI/EIA (2002). EIA-731.1 Systems Engineering Capability Model.

Auyang, S. Y. (2004). Engineering – An Endless Frontier. Cambridge, MA, Harvard University Press.

Babbie, E. (2004). The Practice of Social Research, Wadsworth.

Baik, J. (2000). The Effect of CASE Tools on Software Development Effort. Unpublished Dissertation, USC Computer Science Department.

Baik, J., Boehm, B. W. and Steece, B. (2002). "Disaggregating and Calibrating the CASE Tool Variable in COCOMO II." IEEE Transactions on Software Engineering **Vol. 28**(No. 11).

Berne, E. (1964). Games People Play: The Psychology of Human Relationships. New York, Grove Press, Inc.

Blanchard, B. S. and Fabrycky, W. J. (1998). Systems Engineering and Analysis, Prentice Hall.

Boehm, B. W. (1981). Software Engineering Economics, Prentice-Hall.

Boehm, B. W. (1994). "Integrating Software Engineering and Systems Engineering." The Journal of NCOSE **Vol. 1** (No. 1): pp. 147-151.

Boehm, B. W., Egyed, A. and Abts, C. (1998). Proceedings Focused Workshop #10: Software Engineering and System Engineering, Center for Software Engineering.

Boehm, B. W. and Port, D. (1999). "Escaping the Software Tar Pit: Model Clashes and How to Avoid Them." ACM Software Engineering Notes.

Boehm, B. W., Abts, C., Brown, A. W., Chulani, S., Clark, B., Horowitz, E., Madachy, R., Reifer, D. J. and Steece, B. (2000). Software Cost Estimation With COCOMO II, Prentice Hall.

Boehm, B. W. and Hansen, W. (2001). "The Spiral Model as a Tool for Evolutionary Acquisition." CrossTalk: pp. 4-9.

Boehm, B. W., Reifer, D. J. and Valerdi, R. (2003). COSYSMO-IP: A Systems Engineering Cost Model. 1st Annual Conference on Systems Integration, Hoboken, NJ.

Boehm, B. W., (2003) "Value-Based Software Engineering." ACM Software Engineering Notes.

Boehm, B., Valerdi, R., Lane, J., Brown, A. W., (2005) "COCOMO Suite Methodology and Evolution," CrossTalk - The Journal of Defense Software Engineering.

Brooks, F. P. (1995). The Mythical Man-Month: Essays on Software Engineering, Addison-Wesley.

Chulani, S., Boehm, B. W. and Steece, B. (1999). "Bayesian Analysis of Empirical Software Engineering Cost Models." IEEE Transactions on Software Engineering: pp. 513-583.

Clark, B. K. (1997). The Effects of Software Process Maturity on Software Development Effort. Unpublished Dissertation, USC Computer Science Department.

CMMI (2002). Capability Maturity Model Integration - CMMI-SE/SW/IPPD/SS, V1.1. Pittsburg, PA, Carnegie Mellon - Software Engineering Institute.

Conte, S. D., Dunsmore, H. E., Shen, V. Y., (1986) Software Engineering Metrics and Models, Benjamin/Cummings Publishing Company.

Cook, D. and Weisberg, S. (1999). Applied Regression Including Computing and Graphics, John Wiley & Sons.

Dalkey, N. (1969). The Delphi Method: An Experimental Study of Group Opinion, RAND Corporation.

Ernstoff, M. and Vincenzini, I. (1999). Guide to Products of System Engineering. International Council on Systems Engineering, Las Vegas, NV.

Faisandier, A, Lake, J., Harmonization of Systems and Software Engineering, INCOSE INSIGHT, Vol. 7, Issue 3., October 2004.

Ferens, D. (1999). Parametric Estimating - Past, Present, and Future. 19th PRICE European Symposium.

Freud, S. (1924). A General Introduction to Psycho-Analysis. New York, Washington Square Press, Inc.

Friedman, G. (2003). ISE541 Lecture Notes. USC Industrial & Systems Engineering.

GAO-03-1073 (2003). Defense Acquisitions Improvements Needed in Space Systems Acquisition Management Policy.

Griffiths, W. E., Hill, R. C. and Judge, G. G. (1993). Learning and Practicing Econometrics, John Wiley & Sons, Inc.

Honour, E. C. (2002). Toward an Understanding of the Value of Systems Engineering. First Annual Conference on Systems Integration, Hoboken, NJ.

Horowitz, B. (2004). Systems Engineering Estimation Rules of Thumb. R. Valerdi. Charlottesville, VA: Personal communication.

Humel, P. (2003). Systems Engineering Revitalization at Space and Missile Systems Center. INCOSE Los Angeles Chapter Meeting, The Aerospace Corporation.

Isaac, S. and Michael, W. B. (1997). Handbook in Research and Evaluation. San Diego, CA, EdITS.

ISO/IEC (2002). ISO/IEC 15288:2002(E) Systems Engineering - System Life Cycle Processes.

Jackson, S. (2002). ISE541 Lecture Notes. USC Industrial & Systems Engineering.

Jarvenpaa, S. L., G. W. Dickson, G. W. and DeSanctis, G. (1985). "Methodological Issues in Experimental IS Research: Experiences and Recommendations." MIS Quarterly Vol. 9(No. 2).

Kemerer, C. (1987). "An Empirical Validation of Software Cost Estimation Models." Communications of the ACM Vol. 30(No. 5): pp. 416-429.

Klein, H. K. and Myers, M. D. (1999). "A Set of Principles for Conducting and Evaluating Interpretive Field Studies in Information Systems." MIS Quarterly Vol. 23 (No. 1).

MIL-STD 490-A (1985) Specification Practices.

MIL-STD-499A (1969). Engineering Management.

NASA (2002). Cost Estimating Handbook.

Pandikow, A. and Törne, A. (2001). Integrating Modern Software Engineering and Systems Engineering Specification Techniques. 14th International Conference on Software & Systems Engineering and their Applications.

PRICE-H (2002). Your Guide to PRICE-H: Estimating Cost and Schedule of Hardware Development and Production. Mt. Laurel, NJ, PRICE Systems, LLC.

Rechtin, E. (1991). Systems Architecting: Creating & Building Complex Systems, Prentice Hall.

Rechtin, E. (1998). System and Software Architecture. Proceedings Focused Workshop #10: Software Engineering and System Engineering.

Sage, A. P. (1992). Systems Engineering, John Wiley & Sons, Inc.

Sheard, S. (1997). The Frameworks Quagmire, A Brief Look. Proceedings for the Seventh Annual Symposium of the International Council on Systems Engineering, Los Angeles, CA.

Shrum, S. (2000). "Choosing a CMMI Model Representation." CrossTalk.

Smith, J. (2004). An Alternative to Technology Readiness Levels for Non-Developmental Item (NDI) Software - CMU/SEI-2004-TR-013. Pittsburg, PA, Carnegie Mellon - Software Engineering Institute.

Snee, R. (1977). "Validation of Regression Models: Methods and Examples." Technometrics **Vol. 19** (No. 4).

Sudman, S. and Bradburn, N. M. (1982). Asking Questions: A Practical Guide to Questionnaire Design. San Francisco, CA, Jossey-Bass.

USCM (2002). USCM8 Knowledge Management System, Tecolote Research, Inc.

Valerdi, R., Ernstoff, M., Mohlman, P., Reifer, D. and Stump, E. (2003). Systems Engineering Sizing in the Age of Acquisition Reform. 18th Annual Forum on COCOMO and Software Cost Modeling, Los Angeles, CA.

Valerdi, R. and Kohl, R. (2004). An Approach to Technology Risk Management. 1st Annual MIT Engineering Systems Division Symposium, Cambridge, MA.

Valerdi, R., Rieff, J. E., Roedler, G. J. and Wheaton, M. J. (2004). Lessons Learned From Collecting Systems Engineering Data. Ground Systems Architecture Workshop, El Segundo, CA.

Appendix A: ANSI/EIA 632 Activities

Fundamental Processes	Process Categories	Activities
Acquisition and Supply	Supply Process	(1) Product Supply
	Acquisition Process	(2) Product Acquisition, (3) Supplier Performance
Technical Management	Planning Process	(4) Process Implementation Strategy, (5) Technical Effort Definition, (6) Schedule and Organization, (7) Technical Plans, (8)Work Directives
	Assessment Process	(9) Progress Against Plans and Schedules, (10) Progress Against Requirements, (11) Technical Reviews
	Control Process	(12) Outcomes Management, (13) Information Dissemination
System Design	Requirements Definition Process	(14) Acquirer Requirements, (15) Other Stakeholder Requirements, (16) System Technical Requirements
	Solution Definition Process	(17) Logical Solution Representations, (18) Physical Solution Representations, (19) Specified Requirements
Product Realization	Implementation Process	(20) Implementation
	Transition to Use Process	(21) Transition to use
Technical Evaluation	Systems Analysis Process	(22) Effectiveness Analysis, (23) Tradeoff Analysis, (24) Risk Analysis
	Requirements Validation Process	(25) Requirement Statements Validation, (26) Acquirer Requirements, (27) Other Stakeholder Requirements, (28) System Technical Requirements, (29) Logical Solution Representations
	System Verification Process	(30) Design Solution Verification, (31) End Product Verification, (32) Enabling Product Readiness
	End Products Validation Process	(33) End products validation

Appendix B: Systems Engineering Effort Profile

EIA 632 Fundamental Process	Average	Standard Deviation
Acquisition & Supply	7%	3.5
Technical Management	17%	4.5
System Design	30%	6.1
Product Realization	15%	8.7
Technical Evaluation	31%	8.7

	Conceptualize	Develop	Operational Test & Eval.	Transition to Operation	Operate, Maintain, or Enhance	Replace or Dismantle
Acquisition and Supply	28 (12.3)	51 (18.6)	13 (11.3)	8 (5.0)		
Technical Management	22 (10.0)	38 (9.9)	25 (7.4)	15 (6.4)		
System Design	34 (12.4)	40 (19.4)	17 (9.6)	9 (6.2)		
Product Realization	13 (14.1)	30 (24.3)	32 (16.0)	25 (20.4)		
Technical Evaluation	18 (11.4)	27 (11.0)	40 (17.7)	15 (8.5)		

In each cell: Average (Standard Deviation)

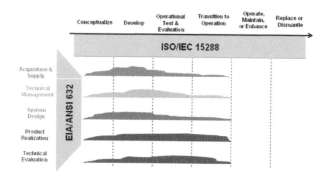

100

Appendix C: List of Industry participants

Aerospace Corporation	Lockheed Martin	SAIC
Abe Santiago	James Evans	Michael McBride
Marilee Wheaton	Carl Newman	Tony Jordano
Paul Mohlman	Rocky Hudson	Don Greenlee
Anh Tu	Garry Roedler	Robert Kaufman
Susan Ruth	Gary Hafen	Ali Nikolai
Paul Stelling	David Lindstrom	Phill Rowell
Darryl Webb	Jeffrey Shupp	Charles Zumba
Pat Maloney	Craig Hayenga	Dick Stutzke
Karen Owens	Greg Kaszuba	
James Horejsi	Keith Young	Softstar Systems
Harlan Bittner	Rick Edison	Dan Ligett
	Paul Robitaille	
BAE Systems	George Walther	SSCI
Jim Cain	Trish Persson	Chris Miller
Donovan Dockery	Bob Beckley	Sarah Sheard
Merrill Palmer	John Gaffney	Jim Armstrong
Gan Wang		
	Northrop Grumman	UC Irvine
Boeing	Linda Brooks	Arnie Goodman
Maurie Hartman	Steven Wong	
Scott Jackson	Albert Cox	US ARMY
	Gregory DiBenedetto	Cheryl Jones
Galorath	Jim VanGaasbeek	
Denton Tarbet		
Evin Stump	Raytheon	
	Gary Thomas	
General Dynamics	John Rieff	
Paul Frenz	Deke Dunlap	
Fran Marzotto	Randy Case	
Sheri Molineaux	John McDonald	
	Bob Vojtech	
Honourcode	Larry Kleckner	
Eric Honour	Greg Cahill	
	Deke Dunlap	
	Ron Townsend	
	Stan Parson	
	Michael Ernstoff	

Appendix D: List of Data Sources

Raytheon	Intelligence & Information Systems (Garland, TX)
Northrop Grumman	Mission Systems (Redondo Beach, CA)
Lockheed Martin	Transportation & Security Solutions (Rockville, MD) Integrated Systems & Solutions (Valley Forge, PA) Systems Integration (Owego, NY) Aeronautics (Marietta, GA) Maritime Systems & Sensors (Manassas, VA; Baltimore, MD; Syracuse, NY)
General Dynamics	Maritime Digital Systems/ Advanced Information Systems (Pittsfield, MA) Surveillance & Reconnaissance Systems/ Advanced Information Systems (Bloomington, MN)
BAE Systems	National Security Solutions/ Integrated Solutions Sector (San Diego, CA) Information & Electronic Warfare Systems (Nashua, NH)
SAIC	Army Transformation (Orlando, FL) Integrated Data Solutions & Analysis (McLean, VA)

Appendix E: Example Estimate Using COSYSMO

Example of the model estimate. The Bayesian calibrated version of the model can be used to estimate systems engineering effort.

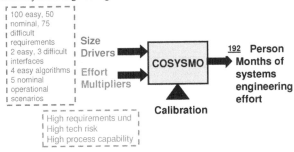

For the given size parameters and three effort multipliers the estimate for systems engineering effort is 192 person months. After a COSYSMO estimate is performed it must go through the common sense test. Does a system with that many requirements, interfaces, algorithms, operational scenarios; and an organization with high requirements understanding and high process capability; and a product with high technology risk warrant 36 person months of systems engineering? If this does not make sense, then one or both of the following two alternatives must be exercised. First, the technical parameters may need to be adjusted to better reflect the system characteristics. Second, the model calibration factor, size driver weights, and cost driver ratings must be adjusted to reflect that organization's productivity.

The person month estimate provided by COSYSMO should be taken into perspective relative to the project dynamics that are present. Costs estimates generally fall into three different areas:

1. *Could Cost*: The lowest reasonable cost estimate involved in fulfilling the essential systems engineering functions.
2. *Should Cost*: The most likely cost involved in providing the systems engineering deliverables to bring a system into a condition of operational readiness.
3. *Would cost*: The highest cost estimated for the systems engineering effort that might have to be performed in order to bring a system into a condition of operational readiness if significant difficulties arise.

These categories, originally provided by (Sage 1992) have been modified to fit this circumstance. What the user is interested in is neither of these, rather, the *Will Cost* is the desired result.

Appendix F: Cost Driver Correlation Matrix

```
REQ    1.0000
INTF   0.5265   1.0000
ALG    0.5141   0.8456   1.0000
OPSC   0.4903   0.2561   0.0655   1.0000
RQMT   0.1283   0.2982   0.3226  -0.0635   1.0000
ARCH   0.1980   0.3288   0.3700  -0.2212   0.4903   1.0000
LSVC  -0.1476  -0.0875  -0.1776   0.0323   0.1697  -0.0055   1.0000
MIGR   0.4756   0.4400   0.4987  -0.0890   0.0369   0.1836  -0.0967   1.0000
TRSK  -0.0123  -0.0433   0.0068  -0.1279  -0.3210   0.0503  -0.1946   0.2932   1.0000
DOCU   0.1177   0.3480   0.2148   0.1774   0.5145   0.1937   0.2437  -0.0370  -0.3781
INST   0.2446   0.3289   0.3275   0.0914   0.4816   0.0886  -0.0967   0.0931  -0.3534
RECU   0.3126   0.2916   0.0163   0.2199   0.0978  -0.0119   0.1256  -0.0224  -0.4047
TEAM  -0.0423   0.0519   0.0464   0.1121   0.2470   0.1283  -0.2207  -0.2857  -0.0372
PCAP  -0.0399  -0.0801  -0.0490  -0.2764   0.4110   0.2430  -0.2874   0.0007  -0.0989
PEXP   0.2582   0.2313   0.2935  -0.0880   0.1273   0.2899  -0.0979   0.4698   0.4012
PROC  -0.0915  -0.0147  -0.1761   0.1627  -0.0804  -0.2814   0.0872  -0.0502  -0.2486
SITE   0.1896   0.3078   0.2076   0.0715   0.3475   0.1806  -0.0660  -0.0010  -0.1088
TOOL  -0.1709  -0.1145  -0.0021  -0.2644   0.2775   0.2287  -0.1376   0.0180  -0.2076
          REQ     INTF      ALG     OPSC     RQMT     ARCH     LSVC     MIGR     TRSK
```

```
DOCU   1.0000
INST   0.5338   1.0000
RECU   0.5162   0.3109   1.0000
TEAM  -0.1476  -0.0064  -0.2477   1.0000
PCAP  -0.0899   0.1028   0.0533   0.3073   1.0000
PEXP  -0.2354  -0.2403  -0.0771   0.0706   0.3159   1.0000
PROC   0.0521  -0.1205   0.2254   0.0044  -0.3552  -0.1893   1.0000
SITE   0.2878   0.3034   0.1877   0.2034  -0.1179   0.0132   0.3818   1.0000
TOOL  -0.0805   0.1631  -0.1848   0.1483   0.1544  -0.0455   0.0905   0.2417   1.0000
          DOCU     INST     RECU     TEAM     PCAP     PEXP     PROC     SITE     TOOL
```

Appendix G: Cost Driver Distributions

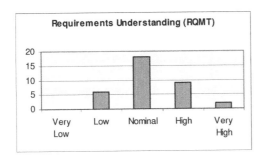

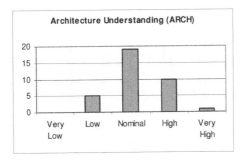

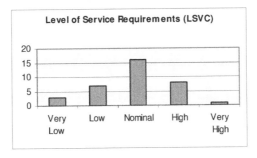

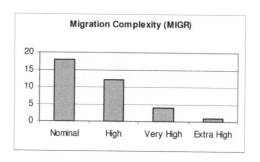

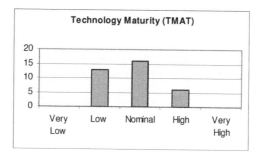

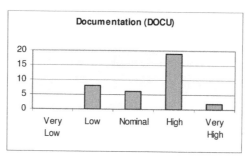

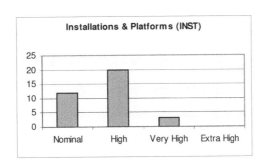

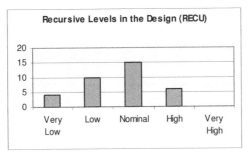

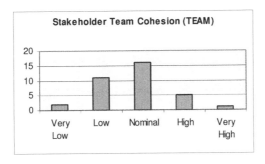

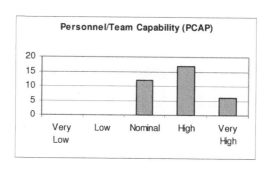

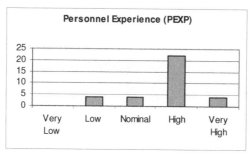

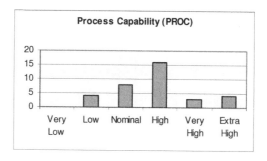

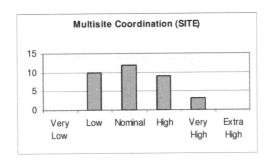

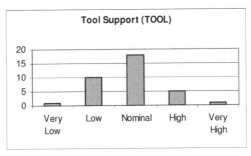

Appendix H: Regression Results for Final Model

```
Data set = COSYSMO, Name of Fit = L1
Deleted cases are
(8)
Normal Regression
Kernel mean function = Identity
Response     = log[SE_HRS_ADJ]
Terms        = (log[SIZE] log[_3COMPLEXITY] log[_3ENVIRONMENT]
log[_3OPERATIONS] log[_3PEOPLE] log[_3UNDERSTANDING])
Coefficient Estimates
Label                    Estimate        Std. Error    t-value    p-value
Constant                 3.65195         0.740909      4.929      0.0000
log[SIZE]                0.820202        0.108061      7.590      0.0000
log[_3COMPLEXITY]0.584024                0.470309      1.242      0.2250
log[_3ENVIRONMENT]0.400851               0.704392      0.569      0.5740
log[_3OPERATIONS]0.956473                0.629541      1.519      0.1403
log[_3PEOPLE]           -2.13820         0.809700     -2.641      0.0136
log[_3UNDERSTANDING]0.0301342            0.269532      0.112      0.9118

R Squared:                     0.819548
Sigma hat:                     0.695759
Number of cases:                     42
Number of cases used:                34
Degrees of freedom:                  27

Summary Analysis of Variance Table
Source          df      SS         MS          F      p-value
Regression       6    59.36      9.89333     20.44    0.0000
Residual        27    13.0702    0.484081
```